J.LEGA

Principles of Map Design

Principles of
MAP DESIGN

Judith A. Tyner

THE GUILFORD PRESS
New York London

© 2010 The Guilford Press
A Division of Guilford Publications, Inc.
72 Spring Street, New York, NY 10012
www.guilford.com

Printed in the United States of America

This book is printed on acid-free paper.

Last digit is print number: 9 8 7 6 5 4 3 2 1

Library of Congress Cataloging-in-Publication Data

Tyner, Judith A.
    Principles of map design / Judith A. Tyner.
        p. cm.
    Includes bibliographical references and index.
    ISBN 978-1-60623-544-7 (hardcover)
    1. Cartography. 2. Thematic maps. I. Title.
    GA105.3.T97 2010
    526—dc22

                                                    2009049691

Maps and Graphics by Gerald E. Tyner, PhD
GIS Consultant James A. Woods, MA

To my mentors

Richard Dahlberg
Gerard Foster
Norman J. W. Thrower

Preface

A map says to you, "Read me carefully, follow me closely, doubt
me not." It says, "I am the earth in the palm of your hand.
Without me, you are alone and lost."

—BERYL MARKHAM, *West with the
Night*, 1942

An earlier version of this book was published in 1992. In the years between its writing and the present version, changes in mapmaking have been enormous. We have
moved in the last 20 years from pen-and-ink drafting to computerized mapping.
Mapmaking is in the midst of a revolution that had its beginnings over 50 years ago.
This revolution is based on changes in technology, in kinds of data, and in social
influences. Data that would not have been available in 1950, such as satellite imagery,
are now routinely available to anyone with Internet access. The Internet itself is a
product of only the last 20 years. Mapmakers have become more aware of the impact
of their products on society and have an increased concern with ethics and privacy.
Technological advances including satellites and computers have had a major impact
on the field. The impact of research on how maps work, how readers perceive maps
and symbols, and visualization has changed our thinking about maps. Rapid changes
in software and hardware continue unabated. A sophisticated cartography lab hardly
more than 15 years ago would have had perhaps 10 desktop computers with "line"
printers, digitizers, and perhaps a plotter; this seems primitive today. GIS exploded
onto the scene in the 1990s (although its antecedents go back to the 1930s). It seems,
in fact, that the only constant in the field is change.

However, if one looks beyond the technology, there are principles that remain
sound regardless of production methods. These principles are the basis of "good"
maps whether produced with pen and ink or the most recent GIS package, whether
printed or viewed online.

It is important to remember also that creating maps goes beyond the look of the page. Maps have an impact on society; they are used in decision making at many levels, from a simple "How do I get there?" to "Where should the money be allocated?" The mapmaker must take into account the purpose of the map, the intended audience, and where and how the map might be used. The mapmaker must never lose sight of the power that maps have.

This book is divided into five parts. Part I is titled Map Design. This may seem contrary to common sense. After all, one must gather data, then select a scale, a projection, and symbols; shouldn't all this come before design? Map design is actually a twofold process. This book focuses on "design" in the broad sense of planning the map, not merely on layout and how to make the map "pretty." Design is a decision-making process and, for maps, includes choosing data, choosing projection, choosing scale, establishing a hierarchy, choosing symbols, choosing colors, and choosing type in order to make an effective map for a given purpose. Thus, design is the heart of mapmaking. Part II focuses on the geographic and cartographic framework. This includes compilation, generalization, projections, and scale. Part III involves symbolization and how to represent various kinds of data. Symbols are often called the "language of maps" and while this isn't strictly true, choice of symbol is critical in the effectiveness of a map. Part IV concentrates on what might be considered nontraditional mapping and more advanced visualization techniques. Here, design principles for web mapping, animated maps, cartograms, interactive maps, and maps for the visually impaired are discussed. Part V, Critique of Maps, is a series of map "makeovers," evaluating and improving maps.

A list of suggested readings is included at the end of each chapter for the reader who would like more information on the material in that chapter, and a complete bibliography that includes the readings plus other sources used in creating the book is provided at the end of the book.

Three appendices are included: a table of common projections, a list of resources, and a glossary of terms. URLs are listed under "Resources" in Appendix B. Those included are primarily government sites such as the U.S. Geological Survey (USGS), the Census Bureau, and cartographic organizations. Few individual websites are included, since they are subject to rapid change and often disappear.

This book does not focus on any specific software, but on principles of making maps. It is not a "how-to" book. Numerous manuals are available for use with different software packages; some of these are listed in the bibliography. The industry-standard software at the time of this book's writing could well be out of date by the time of publication. The principles are those that are generally accepted.

It is the task and objective of a textbook author to translate and summarize current thinking and practices in the discipline. Any textbook is somewhat idiosyncratic and reflects the thinking of the author or authors. It reflects what the author believes is important in the discipline. This book is no exception. I have drawn on many sources, including conversations and input from other mapmakers, and I have tried to present the most accepted principles at the time of writing, but this book is essentially my view of cartography, and any errors that may have insinuated themselves into the text are mine.

ACKNOWLEDGMENTS

No book of this nature is a solo production, and I would like to thank those who helped me along the way. First I would like to thank my three mentors, without whom my career and ultimately this book would not have been possible: Richard Dahlberg, who introduced me to cartography, took the time to answer many "off-the-wall" questions from an eager undergraduate, and encouraged my research interests; Gerard Foster, from whom I learned about teaching cartography; and, finally, Norman J. W. Thrower, my mentor and friend for more years than either of us want to count.

Next are my colleagues at CSU Long Beach—Christopher Lee, Suzanne Wechsler, and Christine Rodrigue, who dug up maps and references and acted as sounding boards; Greg Armento, the Geography and Map Librarian, who let me stash a shelf of cartography journals at home while the library was being remodeled; Mike McDaniel, who read an early draft of the manuscript and made helpful comments—and Nancy Yoho, former student and vice president of Thomas Brothers/Rand McNally, who has been helpful for many years and arranged for tours of the company for my classes, where I always learned as much or more than the students.

The book could not have been completed without Gerald E. Tyner, who took my ideas and sketches and turned them into readable maps and diagrams, and James "Woody" Woods, who fielded arcane GIS problems.

Of course, I thank my family for their assistance and patience in listening to me as I talked out chapters: my son James A. Tyner, of the Geography Department at Kent State University, who was always ready to discuss writing and geography; my son David A. Tyner, a graphic designer, with whom I discussed (and argued) design issues; and my husband, Gerald, who in addition to creating the maps and reading drafts, has supported my research and writing for these many years. Rocky, Punkin, Max, and Bandit—without your "help" the book would have been finished sooner.

Kristal Hawkins and William Meyer of The Guilford Press deserve special thanks for having faith in this book and patiently seeing it through the long process to publication.

Finally, I thank the 1,500 undergraduate and graduate cartography students I have taught through the years. I learned from your mistakes—you taught the teacher.

Contents

Part IV. Nontraditional Mapping

Part V. Critique of Maps

Appendices

PART I

MAP DESIGN

Chapter 1

Introduction

Far from being an antique craft belonging to a bygone era, cartography is the art of geovisualization; a way of sharing spatial knowledge and empowering people through the application of good design, whether the medium is electronic or paper, permanent or perishable, static or dynamic.

—ALEXANDER J. KENT, *Bulletin of the Society of Cartographers* (2008)

THE SCOPE OF CARTOGRAPHY

Who Is a Mapmaker?

The short answer is everyone. We sketch maps on a piece of paper to show how to get to our house, we download maps from the Web and annotate them, we sometimes take a pen or pencil and make a more formal map of a route or a place. Artists make maps for books or magazines and use maps symbolically in their work. It is a mark of the major changes in the field that now almost anyone can make a professional-looking map on the computer. We create maps of data from some spreadsheet programs. Illustration programs allow for more elegant maps to our house than the pencil sketch. Mapping programs and geographic information systems are increasingly affordable and available to the general public. Of course, there are also professionals who have been trained in mapmaking and make their living creating maps.

Cartography, GIS, Visualization, and Mapmaking

These terms are all used to describe the process of making maps. However, they are not synonymous. *Cartography* has been defined by the International Cartographic Association as "the art, science and technology of making maps, together with their study as scientific documents and works of art." It has also been defined as "the production—including design, compilation, construction, projection, reproduction, use, and distribution—of maps" (Thrower, 2008, p. 250).

The term *geographic cartography* is frequently used to distinguish the kinds of maps that geographers use in world and regional studies to distinguish it from *engineering cartography*, which is used for the type of maps that city engineers create for water lines, sewer lines, gas lines, and the like that would be used in planning and engineering. Many of the principles apply to both; the difference is one of scale.

GIS stands for *geographic information systems*, but the "S" is increasingly being used to stand for *science* and *studies* as well. *Geographic Information Science*, and *Geographic Information Studies* are used increasingly. No universally agreed-upon definition has been put forth. Surprisingly, a number of GIS texts do not even attempt to define the term. For our purposes, the following definition, which is the most common, will be used: *A computer-based system for collecting, managing, analyzing, modeling, and presenting geographic data for a wide range of applications.* Geographic information science, then, is the discipline that studies and uses a GIS as a tool. GIS is not simply creating maps with a computer. The technology is a very powerful tool for analyzing spatial data; while maps can be and are produced with GIS, their main power is analytical. GI scientists do not consider themselves primarily as mapmakers. Although they may produce maps as an end product, their primary emphasis is on *analysis* of the data. In fact, it is comparatively recently that GI systems people have given much thought to *presentation* of data. The types of symbolic representation have been limited as well, but a major recent thrust has been creating new symbol types that would be difficult or impossible to do without computer assistance.

Mapmaking is a generic term that refers to creating maps by any method whether manually or by computer regardless of purpose or scale.

In recent years, since the introduction of GIS, there has been debate over the relevance of cartography. This debate is usually caused by misunderstanding of the terms and the history of GIS. When computers were first introduced into mapmaking, and classes were offered, university departments often made a distinction between cartography classes that utilized manual methods of pen-and-ink drafting and "computer cartography" which utilized rudimentary mapping software and CADD (computer-assisted design and drafting) programs. Eventually the computer cartography classes became GIS classes and all cartography classes utilized the computer with GIS software and perhaps illustration/presentation software, but many people continued to assume that cartography was a manual skill or one that was concerned strictly with the layout of map elements and typography.

Visualization or *geovisualization* also has no agreed-upon definition. Some identify visualization as a "private" activity that involves exploring data to determine relationships and patterns of spatial data. The ESRI corporation defines visualization as "the representation of data in a viewable medium or format." Commonly, definitions of visualization include reference to computer technologies and interactive maps. Two models have been proposed to explain visualization and communication and have gained wide acceptance. Both distinguish between visual thinking and visual communication as "private" activities and "public" activities. DiBiase's model (Figure 1.1) distinguishes between visual thinking and visual communication, with visual thinking being concerned with exploration of data and visual communication being concerned with presenting data. MacEachren's model of visualization and communication (Figure 1.2), often called the "cartography cubed diagram," describes

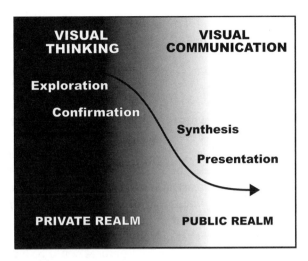

FIGURE 1.1. DiBiase's model distinguishes between visual thinking and visual communication. From DiBiase, David (1990). Reprinted by permission.

visualization as private, interactive, and revealing unknowns, while communication is public, noninteractive, and revealing knowns. Although these models are generally considered the best explanations of visualization, presentations of animated maps and "flythroughs" are often described as visualizations. In this book we will be concerned with both the private visualizations that occur at the planning stage and the public communication or presentation that occurs when the map is published or put online.

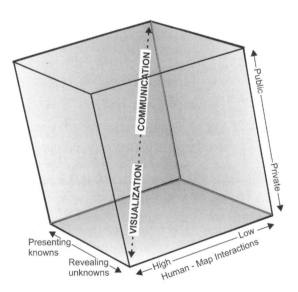

FIGURE 1.2. MacEachren's "cartography cubed" model. From MacEachren, Alan M. (1994b). Reprinted by permission.

In this book the terms *cartography* and *mapmaking* will be used for creating maps of any type by any method, and GIS will be used when a dedicated GIS is required.

What *Is* a Map?

Surprisingly, this is a question for which there is no easy answer. We all "know" what a map is, but that definition can vary from person to person and culture to culture. A general definition from 40 years ago was, "A graphic representation of all or a part of the earth's surface drawn to scale upon a plane." However, questions arose. What about the moon and other extraterrestrial features? If it looks like a map but lacks an indication of its scale, is it a map? Can an annotated satellite image (one with names of features printed on it) be considered a map? Is a globe a map? What about 3-D representations? Purists would say that a "map" with no scale is a diagram and that 3-D representations are models. The moon and planets could be handled by inserting "or other celestial body" into the basic definition.

But then one finds that some non-Western cultures have made representations of place that do not fit the "official" definition, but still function as maps. Navajo sand paintings, Australian Aboriginal bark paintings, and Marshall Island stick charts (Figure 1.3) all function as maps. There are also oral maps, mental maps, and performance maps. Where do these fit in the definition?

J. H. Andrews compiled a list of 321 definitions of "map" made from 1694 to 1996 (Andrews, 2009). Obviously, this is a subject that can be, and is, debated endlessly in seminars, conferences, and over coffee. It is easy to be flippant, but the definition of map sometimes determines what is "worthy" of study by cartographers.

For the purpose of this book I will use a functional definition of a map— that is, if it has a map function, it is a map—and I will define *map* as "a graphic representa-

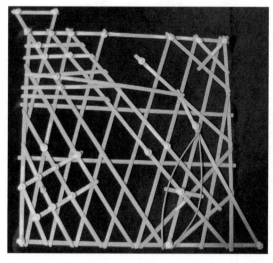

FIGURE 1.3. Marshall Islands stick chart. Three types were made: local, regional, and instructional. Shells represent islands, palm ribs illustrate currents and wave patterns. Author photograph.

tion that shows spatial relationships." In this book we will not discuss designing sand paintings or stick charts; I will confine the discussion to flat maps that show spatial relationships, but I will look at maps for the visually impaired and maps for the computer monitor and the Web in addition to those drawn on paper.

Kinds of Maps

Since maps can represent anything that has a spatial component, there are hundreds of possible map types; however, these can be grouped into a few categories. One categorization is based on map function. These functional categories are *general-purpose maps*, *special-purpose maps*, and *thematic maps*. As is common, there is not complete agreement among cartographers about these terms or categories.

General-purpose maps, or *reference maps*, as the name suggests, do not emphasize one type of feature over another. They show a variety of geographic phenomena (political boundaries, transportation lines, cities, rivers, etc.) and present a general picture of an area. They are used for reference, planning, and location. Commonly, the state or regional maps in an atlas are of this type, and topographic maps are often placed in this category.

Special-purpose maps are created for a very specific type of user. Geologic, soil, and cadastral maps are included here. Such maps are usually large scale (showing a small area and much detail), and the user is usually familiar with the subject, if not the area. Navigation maps, which include all types of maps created for route finding, such as aeronautical charts, nautical charts, and road maps, are often included under the special-purpose heading, although some consider them to be a separate map type (Figure 1.4). Special-purpose maps tend to be made at agencies or corporations and by a team of people.

Thematic maps have been called a variety of names (special subject, statistical, distribution, and data maps), but the term "thematic" is now generally accepted. Thematic maps normally feature only a single distribution or relationship, and any other information shown (base data) serves as a spatial background or framework to help locate the distribution being mapped. Thematic maps may be either qualitative or quantitative. That is, they show some characteristic or property, such as land use, or show numerical data, such as temperatures, rainfall, or population (Figure 1.5). In this book the primary emphasis is on thematic maps although the design principles apply to all map types.

Thematic maps were first widely used in the 19th century. These maps are commonly used in atlases as an adjunct to general maps. Thematic maps are the primary map type seen in newspapers, journals, reports, and textbooks.

Purpose of Thematic Maps

Thematic maps can be made to represent almost any phenomenon, visible or invisible. They can show actual features on the earth, such as rivers, mountains, and

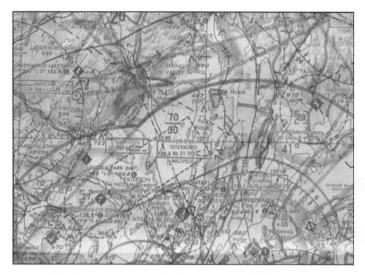

FIGURE 1.4. Maps for navigation, whether for roads, air, or sea, are considered special-purpose maps. This is an aeronautical chart.

roads; conceptual features, such as the earth's grid or county boundaries; and ideas and beliefs, such as locational preference or political ideologies.

Whatever the topic, a thematic map is made for one of three broad purposes: (1) to provide information on what and perhaps how much of something is present in different places, that is, data storage; or (2) to map the characteristics of a geographic phenomenon to reveal its spatial order and organization, that is, visualization; or (3) to present findings to an audience, that is, communication.

Data storage is a map function that has long been recognized, although the term

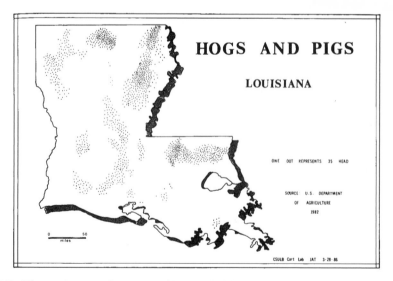

FIGURE 1.5. Thematic map. Courtesy of James A. Tyner.

is recent. On early maps, the data stored were usually locational. Positions of islands, routes, or records of boundaries for the tax collector are examples of this type of early data storage. Maps still perform this function, but the kinds of data stored have expanded and sometimes the method of storage has changed. For example, boundary lines may be recorded and stored in digital form and printed on demand.

Maps are, by their very nature, spatial representations. That is, they show positions in space. They are uniquely suited therefore to portray features of the earth's surface (for terrestrial maps) or to show the *spatial relationships* of features to one another. No other device can do this as well as a map. Text, tables, and even graphs do not possess the spatial component and do not allow readers to see distributional patterns.

Because of their ability to show spatial relationships, maps are used as analytical and explanatory tools. Some geographic patterns cannot be recognized until they are presented in map form; therefore, maps are often made to aid a researcher in identifying or correlating distribution patterns, that is, visualizing data.

Finally, maps are used to present or communicate information to an audience, which might be readers of a report, students studying a textbook, a shopper looking at a "you are here" map of a mall, or a hiker checking a route at a trail head.

Limitations of Maps

In spite of their usefulness, maps have limitations. Many map readers are not aware of these limits (and the appearance of some published maps shows that not even all mapmakers are aware of these limitations). Part of the problem is that people often assume that a map shows everything, like a photograph. A photograph taken from the air from low-flying aircraft shows whatever is in view: houses, streets, cars, the family dog, and even laundry drying in the backyard. Figure 1.6 shows the port of Long Beach with boats and their wakes in addition to docks and buildings.

Maps are not photographs. This seems an obvious, even simplistic, statement, but the distinction is important. Photographs are not selective except through the inherent selection of *resolution*, that is, the size of objects large enough to be seen. This varies with the height of the aircraft or satellite and the capabilities of its sensor. Maps are graphic representations, which by their very nature are selective and symbolic, that is, *generalized*. Maps do not show every bit of available information. To do so would clutter the map with information that isn't relevant to the theme or topic of the map. It would obscure the message. Symbols substitute for images of objects. The map in Figure 1.7 shows the same area as the picture in Figure 1.6.

While selection is vital, it also acts as a major limiting factor on maps. Although some "missing" facts may be inferred from other information, normally one cannot read into a map information that is not shown. For example, one cannot determine the exact nature of terrain from a map that shows the pattern and amounts of rainfall, although some educated guesses can be made.

Selection also involves bias, a subject that has been of interest to researchers in the past 30 years. The decision of what to include on a map depends on many objective factors, but also on subjective factors, such as what the cartographer, the mapping agency, or the client want to show and emphasize. All maps are biased to some extent. This does not mean they are therefore evil or incorrect, but one should

FIGURE 1.6. Aerial photographs are not selective. This photograph of the Port of Long Beach shows boats and their wakes as well as fixed features.

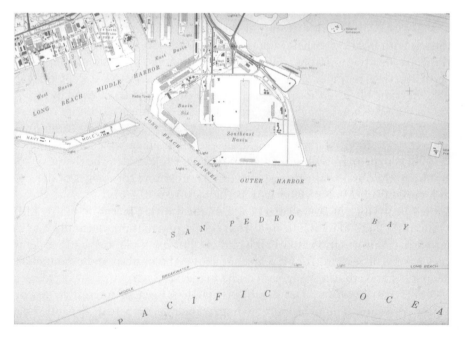

FIGURE 1.7. Maps are generalized and symbolic. This USGS topographic map shows the same area as in Figure 1.6.

be aware of this fact when using or making a map. Selection is discussed more fully in Chapter 5.

A second limitation is imposed by *scale*. In part, maps are selective because of scale considerations. Maps are drawn smaller than reality, and in this process of scaling down, some detail is necessarily lost. The greater the reduction from actual size, the more generalized the information becomes.

A third limiting factor results from the inescapable fact that the earth is spherical and maps are flat. It is not possible to transform the spherical shape of the earth onto a flat map without some distortion somewhere on the map. However, it is possible to minimize distortion or to confine it to a part of the map away from the area of primary interest. The process of transforming the earth's grid to a plane is called *projection*.

Finally, as we have seen, maps are limited to showing *spatial* relationships and characteristics such as distance, direction, position, angle, and area. Maps cannot effectively illustrate ideas and concepts that lack a spatial component. Sometimes a word is worth a thousand pictures. The mapmaker must decide whether a map is the most appropriate medium for communicating an idea.

The mapmaker has a responsibility to the map user to create a map that minimizes the map's limitations or uses them to enhance communication. This subject is treated in detail in later chapters.

The Power of Maps

Maps are powerful tools. They are often accepted at face value and their veracity is seldom questioned by users. Whereas a reader might question the sources of a table, or of a text, he or she often assumes a map to be accurate until it is proven otherwise. This is especially the case with GIS maps because a computer-created map conveys a sense of great accuracy. Blithely assuming maps to be accurate can have serious consequences. Maps are used in decision making, whether deciding which route to take, which county should receive the most money, or where a boundary line should be drawn.

In the worst-case scenario, maps can and do kill. The most famous recent example is the 1998 tragedy of 20 deaths in Italy when a low-flying jet plane cut the cables of a ski gondola that was not shown on the pilots' map. Other examples include deaths by friendly fire from using outdated maps and plane crashes into towers not shown on a map.

Often, the problem is using a map for which it was not designed, such as using a road map to decide election districts. However, a major problem for our purposes is failing to design a map for it's intended purpose. A ship's navigator isn't interested in roads on land, he or she is concerned with possible hazards to navigation. Leaving out a road will cause no problems, but leaving out a submerged hazard can lead to tragedy.

Maps are used in planning, real estate, and government that influence decisions on location of industries and developments, where one buys a home, and where congressional districts are drawn. Maps are also available online that show housing values, taxes paid (or unpaid), the size of the house, the number of rooms, and when the house was last sold. This information is in the public record, but its availability to anyone with an Internet connection anywhere in the world raises concerns. Thus, we must consider the social impact of the maps we make and the ethics of the field. This will be a recurring theme in this book.

The Mapping Process

The mapping process is not linear, but in a book of this sort the material must be presented in a linear fashion. Remember that at times two or more processes are going on at once. Creating a map can be compared to writing a paper, a thesis, or a book. The stages fall into four categories: planning, analysis, presentation, and production/reproduction (Figure 1.8).

In the *planning* phase, the cartographer must have a clear idea of the purpose and topic of the map, where it will be presented, and for whom it is designed. This will govern the type of data collected. *Analysis* involves collecting, synthesizing, and analyzing the data. Data may be gathered in the field, from statistical sources, from other maps, from imagery, or online. Any combination of these data sources may be used. The data are analyzed and symbolized using statistical tools, which may be built into a GIS. For *presentation*, the elements of title, legend, scale, orientation, text, and illustrations are organized into a layout. At this stage, the mapmaker must know where and how the map will be viewed or produced—computer moni-

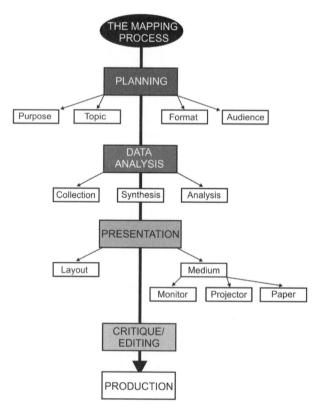

FIGURE 1.8. The mapping process.

tor, printed paper map, Internet. After the map is created, but before *production/ reproduction*, one should critique and edit the map. Are there errors of fact or errors in spelling? Do the symbols, colors, and lines work? Finally, the map is "published." This could be as simple as printing from the computer, making Xerox copies, or posting on the Web, or the map can be sent to a printer and publisher for distribution in thousands of copies.

ANTECEDENTS OF MODERN CARTOGRAPHY

Throughout the history of cartography there have been periods of great change interspersed with periods of quiescence. The periods of change encompass many decades and may be major or minor. Some periods of change are so great that they are described as revolutions. The revolutions are characterized by three factors: technology, data, and social/philosophical changes. One period of major revolution in the Western world was the Renaissance (c. 1350–c. 1650). A primary technological factor was the invention of printing, which lead to a wider dissemination of ideas. Books and maps became available to a greater number of people. European exploration of the western and southern hemispheres provided information, that is, data, that allowed a more accurate and complete representation of the continental outlines. Social and philosophical shifts, including the rediscovery of Ptolemy's works, lead to changes in the nature of maps. The scope of this text does not allow a complete history of cartography, but because events in the 20th century, especially the last half of the century, are important for understanding current theory and practices, this period is summarized here. Some references to more complete histories are provided in the bibliography.

The 20th-Century Revolution

Mapmaking is now in the midst of a major revolution that had its beginnings in the middle of the 20th century. As with any revolution the changes involve technology, increased and new data, and philosophical factors. World War II was a major impetus in that it created a need for up-to-date maps of widespread areas. The number of maps required was huge and they needed to be created rapidly. In the United States, at that time, a call went out for thousands of people to be trained and employed in mapmaking, photogrammetry, and air photo interpretation. After the war, these people, many of them women, continued working in cartography as the government vowed never to be caught short again.

At the end of the war, geography departments began teaching cartography, which had previously been concentrated in civil engineering. They were especially concerned with "geographic cartography" or thematic cartography rather than surveying and mapping or engineering cartography. Geography had, of course, always been involved with maps, and at some periods of time "geographer" was synonymous with "mapmaker" or "cartographer." However, until the 1950s, geographers considered cartography a tool and a skill, not a science or research area, and little research was done on how maps work. There were few textbooks available. Geographical journals published articles on map projections and the history of maps, but little on symbols and nothing on design. This changed after World War II.

In 1938, the primary cartography textbook in the United States was Erwin Raisz's (pronounced Royce) *General Cartography*. Raisz's book emphasized practical aspects and he believed that the lectures of a cartography course should concern history and the laboratory portion should be focused on lettering and the use of drafting instruments.

After the war, a geography graduate student, Arthur Robinson, who had headed the mapping division of the Office of Strategic Services (OSS) during the war, returned to his studies at Ohio State University. His dissertation topic was unusual in that it dealt with map design. The dissertation was published as *The Look of Maps* in 1952 and was considered groundbreaking. It covered such subjects as color, typography, and map structure. Robinson accepted a teaching position at the University of Wisconsin and began a research program that stressed how maps worked. Dissertations carried out under Robinson were often psychophysical studies of symbols such as graduated circles and isopleths. Robinson wrote the primary cartography textbook of the last half of the 20th century, *Elements of Cartography*, which went through six editions from 1953 to 1995.

In the same period, other cartographers who had been involved in mapping during World War II took teaching positions at universities and cartography began to emerge as a discipline. Two of those cartographers, George Jenks, at the University of Kansas, and John Sherman, at the University of Washington, and their students also carried out research on how maps function.

Technology

By the 1960s new technology was revolutionizing the field. Computer programs were being devised that could create maps from digital data. The Harvard Laboratory for Computer Graphics introduced SYMAP in the 1960s. Although the maps were crude, the potential could be seen. In the early days the only printers were line printers that operated as automatic typewriters and all symbols on the map were made up of alphanumeric characters (Figure 1.9). SYMAP maps were of little use for presentation, but they did permit rapid spatial representation and analysis of data.

Another major technological impact was *remote sensing*. Aerial photographs had been widely used during World War II and before, but with the advent of satellites and sensors a wealth of high-resolution imagery became available. We take for granted the satellite imagery displayed on weather reports and we track hurricanes from our living rooms, but this wasn't possible until the last third of the 20th century. This imagery is a part of our mapping data.

The concepts for *geographic information systems* date to the 1930s when geographical analysis was carried out by placing information on a series of clear plastic layers. Modern GIS utilizes virtual layers in analysis.

Philosophical Factors

In the past 50 years our ideas about cartography have changed and new approaches to making and studying maps have appeared.

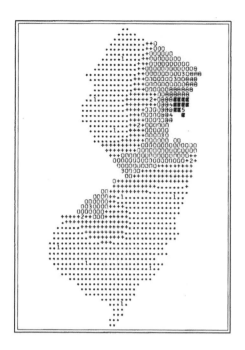

FIGURE 1.9. A map created with SYMAP.

COMMUNICATION THEORY

The research of Robinson, Jenks, Sherman, and their students conducted incorporated ideas coming from *communication theory*. "How do I say what to whom and is it effective?" (Koeman, 1971) was one of the questions they were asking. This thinking was quite different from that of earlier days when cartographers made maps with little concern for how the reader would perceive the map. Instead, in the communication paradigm, the cartographer asked what the reader would get from the map and whether the map would effectively convey the cartographer's message. Individual symbols, such as graduated circles and isopleths, were tested for effectiveness. Today, such research is still being carried out for animated maps and multimedia maps.

VISUALIZATION PARADIGM

By the 1990s some criticized the communication approach to cartography, seeing the methodology as a search for a single optimum representation. Sophisticated computer programs had been developed that permitted interactive exploration of data and the visualization paradigm was introduced. Like GIS, visualization is not an entirely new concept. If we define visualization as a private activity that involves exploring data to discover unknowns, then thematic cartographers have been involved in visualization for a very long time. Early visualizations were not done with the aid of a computer, but with tracing paper and colored pencils while the cartographer examined the data and experimented with representations. With the advent of computers and the rise of scientific visualization, the change was logical.

CRITICAL CARTOGRAPHY

In addition to the research themes of communication and visualization, which pertain directly to creating maps, *critical cartography* has emerged. Cartographers have examined bias on maps for over a century, and during World War II and after studies examined maps as tools for persuasion and propaganda and looked at distortion on maps; seminal works by Brian Harley in the 1980s took this kind of study in a new direction. Harley's "Deconstructing the Map" drew on literary theory and examined maps as texts. This was originally applied primarily to old maps, but critical cartography and the social implications of maps are now a major theme in analyzing modern maps. Denis Wood's *The Power of Maps* explores this theme. Other themes are feminist cartography, maps for empowerment, emotional maps, and the like.

SOCIAL IMPLICATIONS

With the use of GIS in producing maps and their distribution on the Internet, there has been increasing concern with the ethics of the field and the impact that maps have on society. John Pickles's *Ground Truth: Social Implications of GIS* is one of the early studies of this aspect of the field. This is, of course, closely tied to critical cartography.

SUGGESTIONS FOR FURTHER READING

Andrews, J. H. (2009). *Definitions of the word "map," 1649–1996*. Available at *www.usm. maine.edu/~maps/essays/andrews.htm*

Harley, J. Brian. (2001). *The New Nature of Maps: Essays in the History of Cartography* (Paul Laxton, Ed.). Baltimore: Johns Hopkins University Press.

MacEachren, Alan M. (2004). *How Maps Work: Representation, Visualization, and Design*. New York: Guilford Press.

Monmonier, Mark. (1996). *How to Lie with Maps* (2nd ed.). Chicago: University of Chicago Press.

Pickles, John. (Ed.). (1995). *Ground Truth: The Social Implications of Geographic Information Systems*. New York: Guilford Press.

Raisz, Erwin. (1948). *General Cartography* (2nd ed.). New York: McGraw-Hill.

Robinson, Arthur H. (1952). *The Look of Maps: An Examination of Cartographic Design*. Madison: University of Wisconsin Press.

Robinson, Arthur H., et al. (1995). *Elements of Cartography* (6th ed.). New York: Wiley.

Thrower, Norman J. W. (2008). *Maps and Civilization* (3rd ed.). Chicago: University of Chicago Press.

Tyner, Judith. (2005). Elements of Cartography: Tracing 50 Years of Academic Cartography. *Cartographic Perspectives, 51*, 4–13.

Wood, Denis. (1992). *The Power of Maps*. New York: Guilford Press.

A number of GIS texts exist, many of which are aimed at specific applications, such as business, natural sciences, or geography. The reader might find the following general texts useful:

Chang, Kang-tsung. (2006). *Introduction to Geographic Information Systems* (3rd ed.). New York: McGraw-Hill Higher Education.

Davis, David E. (2000). *GIS for Everyone* (2nd ed.). Redlands, CA: ESRI Press.

Harvey, Francis. (2008). *A Primer of GIS*. New York: Guilford Press.

Wade, Tasha, and Sommer, Shelly. (2006). *A to Z GIS: An Illustrated Dictionary of Geographic Information Systems*. Redlands, CA: ESRI Press.

Chapter 2

Planning and Composition

Nothing is more commonplace or easier than making maps.
Nothing is as difficult as making them fairly good. A good
geographer is all the more rare for needing nature and art to be
united in his training.

—JACQUES-NICOLAS BELLIN (1744,
quoted in Mary Pedley, *The
Commerce of Cartography*, 2005)

WHAT IS MAP DESIGN AND WHY DOES IT MATTER?

When we speak of map design there are two meanings: *layout* of design elements and *planning* the map. Layout involves decisions such as "Where should I place the title, where should the legend and scale go?"; in art, this is called *composition*. Design in the sense of planning begins before a single line is drawn and includes deciding what information will be included and choosing a projection, the scale, and the type of symbols. It is at the heart of the map creation process. In this chapter we look at both aspects of design. The remainder of the book will assist you in making design decisions.

Map users form their spatial concepts of a place, in large part, from maps, whether it is a neighborhood, a region, the world, or the universe; maps are used in decision making, as we saw in Chapter 1. The information presented on a map can have far-reaching consequences, a reality that places heavy responsibility on the mapmaker. Objective mapmakers are obligated to make maps as clear and truthful as possible.

At the same time there is considerable leeway for creativity in new approaches and techniques. Otherwise there would be no changes in map design. New technology, whether the rise of lithographic printing in the 19th century (invented 1796) or the use of computers in the 20th introduced changes in designs and symbols on maps.

Design is a holistic process; language is a linear process. Although I can identify certain steps that must be taken in mapmaking, they are not necessarily followed in a specific order, and, in fact, several may be taken simultaneously. However, I cannot, in a book, consider all aspects of design at once, but must break them into steps.

Goals of Design

Any design, whether of maps or buildings, has certain goals: clarity, order, balance, contrast, unity, and harmony. These must be kept in mind when planning a map.

Clarity

A map that is not clear is worthless. Clarity involves examining the objectives of the map, emphasizing the important points, and eliminating anything that does not enhance the map message. Although removing data can be carried to an extreme, as in the case of propaganda maps, putting the names of every river on a population map simply clutters the map and makes the thematic information hard to read (Figure 2.1).

Order

Order refers to the logic of the map. Is there visual clutter or confusion? Are the various elements placed logically? Is the reader's eye led through the map appropriately? Since the map is a synoptic, not a serial, communication, cartographers cannot assume that readers will look first at the title, then at the legend, and so on. Studies of eye movements show there is considerable shifting of view. Rudolph Arnheim has noted that the orientation of shapes seems to exert an attraction because the shape of

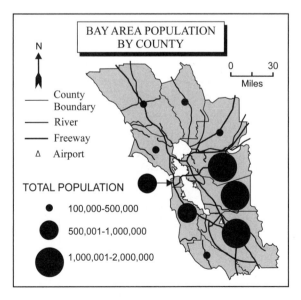

FIGURE 2.1. A map with too much "clutter" is unclear. The rivers, freeways, and airport do not add to the map topic, and, in fact, obscure it.

the elements on a page creates axes that provide direction. That is, vertical lines lead the eye up and down on the map; horizontal lines lead the eye left and right.

Balance

Every element of the map has *visual weight*. These weights should be distributed evenly about the *optical center* of the page, which is a point slightly above the actual center, or the map will appear to be weighted to one side or unstable (Figure 2.2). While this doesn't affect the readability or usefulness of the map, it is a factor in its appearance.

Generally, visual weight within a frame depends on location, size, color, shape, and direction. According to Arnheim (1969, pp. 14–15), visual weights vary as follows:

- Centrally located elements have less weight than those to one side.
- Objects in the upper half appear heavier than those in the lower half.
- Objects on the right side appear heavier than those on the left side.
- Weight appears to increase with increasing distance from the center.
- Isolated elements have more weight than grouped objects.
- Larger elements have greater visual weight.
- Red is heavier than blue.
- Bright colors are heavier than dark.
- Regular shapes seem heavier than irregular shapes.
- Compact shapes have more visual weight than unordered, diffuse shapes.
- Forms with a vertical orientation seem heavier than oblique forms.

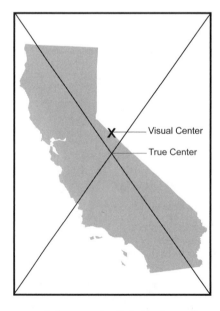

FIGURE 2.2. The visual center of the page is slightly above the actual center.

Closely tied to balance is white space. *White space* is any area within the map frame that is not taken by the map outline itself. A certain amount of white space is required to set the map off and not crowd the page, but usually one should put the largest map possible on the page while still leaving room for other required elements, such as title, legend, and scale. Too often, one sees a small map and the remainder of the page is filled with large north arrows, oversize bar scales, illustrations, and the like that fill the page but overshadow the map (Figures 2.3 and 2.4).

Contrast

A large part of the clarity of the map derives from contrast. *Contrast* is the difference between light and dark, thick and thin, heavy and light. A map created with only one line weight, one font size, and one font lacks contrast, is boring to look at, and is hard to read (Figure 2.5). Some early computer maps lacked contrast because the pen plotters used at the time had only one pen size available; line width could be varied only by cumbersome additional programming steps and commands. Now, of course, sophisticated software is available and today's printers allow a wide variety of fonts and lines so there is no excuse for lack of contrast.

Unity

Unity refers to the interrelationships between map elements. Lettering is not chosen in isolation; it must be legible over any background colors and shades, must not conflict with chosen symbols, and must suit the topic of the map (Figure 2.6). Unity

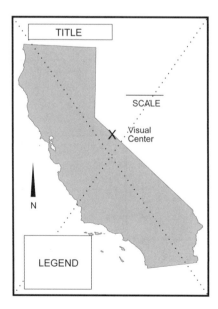

FIGURE 2.3. The layout on the left is poorly balanced. On the right, the page has many elements, but the subject area takes up too little of the available space.

FIGURE 2.4. This is a better layout and use of the available space.

means that the map appears to be a single unit, not a collection of unrelated bits and pieces.

Harmony

Do all of the elements work well together? Do the chosen colors clash? Are patterns jarring to the eye? Do the text fonts complement one another? Does the overall map

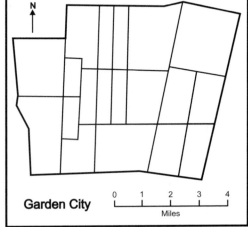

FIGURE 2.5. The figure on the left has no contrast and is bland; the figure on the right has better contrast.

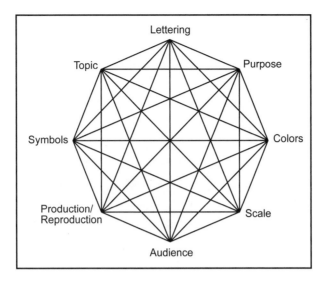

FIGURE 2.6. All of the elements of a map are interdependent.

have a pleasing appearance? While this might not be a problem for a map created for oneself to analyze a geographic problem, if the map is to be presented to a larger audience, it can mean the difference between acceptance of the map and its message or rejection. Simplistically, audiences prefer a pleasing map.

DESIGN AS A PLAN

Formulating the Plan

Design is a decision-making process. Many choices must be made in order to create an effective map whether for visualization or presentation. Before beginning, there are a number of questions to ask. The answers to these questions determine what projections, symbols, scale, colors, type, and all other components will be chosen.

Is a Map the Best Solution to the Problem?

Is a map the best product? There are times when a table or graph might be more appropriate. In general, if the subject has a spatial component, or if spatial relationships are important, then a map is a suitable solution.

What Is the Purpose of the Map?

How will this map be used? Is the map designed to show research findings, to store information, to teach concepts, or to illustrate relationships? The message will probably be unclear unless the cartographer has a definite idea of the purpose of the map. Figure 2.7 shows two maps of the same basic subject designed for different purposes. Note the variations in emphasis.

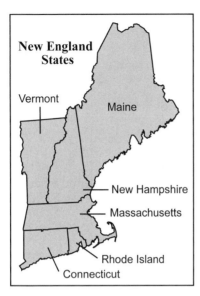 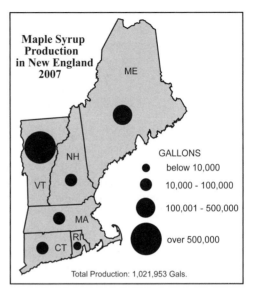

FIGURE 2.7. The varying purposes of these maps is reflected in the design. The map on the left is a simple location map, while the map on the right shows production of a product.

What Is the Subject or Theme of the Map?

A map for navigation has different requirements than a map that simply shows locations or one that shows population density. The theme and location have a bearing on the choice of projection, scale, and degree of generalization. Distribution maps require equal-area projections, a map of wheat distribution does not need a detailed coastline, and midlatitude areas are better represented on conic projections than on cylindricals, for example. Each of these topics is discussed more fully in the relevant chapters.

What Is the Intent of the Map?

Will it explain, will it tell a story, will it be used to persuade, or will it describe? Like writing, maps can be expository, narrative, persuasive, or descriptive. Maps included with research articles are normally used to explain; a map accompanying a story or history may explain or tell a story; a map in a political journal, advertisement, or newspaper may be used to persuade; and some maps simply describe, as in the sense of "you are here." Each of these intentions has somewhat different requirements. Again, these topics will be dealt with in the chapters on color, generalization, and symbolization.

Obviously, purpose, theme, and intent are closely related.

Who Is the Audience?

What are the audience *characteristics*? What is the age of the audience? How familiar with the map subject are they? How map-literate are they? How is their eyesight?

Maps for the visually impaired have different requirements than maps for those with normal vision. Maps for elementary school textbooks have different requirements than maps in scholarly works (Figure 2.8).

What are the *user needs*? How will the readers use the map? Where will they use the map? What are the conditions for reading the map? Will the map be consulted while sitting at a desk, while driving, while on a bicycle tour, or as a reference? These maps will have different requirements because of the needs of the user. A map for a cab driver, which is consulted "on the fly," has different requirements than one intended for a tourist walking along a nature trail.

Too often mapmakers lose sight of their audience. Who is going to use the map and for what should always be at the forefront of the mapmaker's mind whether one is making a map of sewer lines, a newspaper map showing current events, or a map in a textbook. The needs of a city planner, a pilot, and a student are different.

What Is the Format?

Format refers to size and shape of the page or screen and whether color can be used. It ties to where the map will be reproduced. Most professional journals, such as the *Annals* of the Association of American Geographers, *The Professional Geographer*, and *CAGIS*, have a standard format; these standards are available from the editor. Many such journals publish illustration requirements in each issue. When books are designed, the art editor determines the page format. Maps for theses and dissertations have specific formats determined by the university, newspaper maps must conform to column sizes, and maps in business reports will conform to the page size of the printed report. Maps that will be viewed on a monitor or that will be projected onto

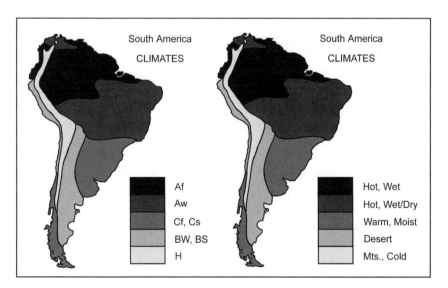

FIGURE 2.8. Map design varies with the audience. The map on the right is for elementary school children and identifies climates with descriptive terms; the map on the left is designed for college age and uses Koeppen climate designations.

a screen have different requirements from printed maps. New formats have become available, such as tiny monitors used on GPS screens, cell phones, and MP3 devices; these have different requirements than wide-screen computer monitors.

Since color is so ubiquitous on monitors and color printers, it is easy to forget that it isn't always an option. Color printing in journals and books is still expensive. Some scholarly journals may require an author to pay for color illustrations. Asking what the format is will save a great deal of grief and reworking. A map designed for color cannot simply be reproduced in black and white. This topic is discussed in Chapter 4.

How Will It Be Produced?

Most maps today are produced with a computer, although some maps are still hand-drawn. In some cases maps are drawn by hand because of lack of computer access; in other cases, such as maps for book illustrations, it is artist preference. The principles of design apply whether the map is drawn with pen and ink or a sophisticated computer, but one should have an idea of how the map will be made at the beginning.

Software for computer-produced maps is of four types: GIS, illustration/presentation, CADD (computer-assisted design and drafting), mapping, or some combination of these. GIS software is a powerful analytical tool with map presentation capabilities. With GIS, data can be linked to places and calculations can be made. As of this writing there are some design limitations and some types of symbol that are difficult or impossible to create using GIS. These problems will be solved at some point. By the same token, some symbols that are easy to produce with GIS cannot easily be created manually or with presentation software. Presentation or illustration software, such as Adobe Illustrator or CorelDraw, is used by graphic artists and allows for highly creative products. However, such software does not allow analysis, calculation, or linking of data to locations automatically. If these capabilities are not needed, a presentation program can be a good choice. Like illustration programs, CADD doesn't allow for analysis. There are some mapping programs, such as Microsoft MapPoint, that have limited GIS capability and allow simple analysis and creation of maps, but do not allow much flexibility in design and composition. Some recent mapping programs, such as Ortelius and Map Publisher, combine GIS and design (Figure 2.9). If one is using a dedicated GIS, combining it with a presentation program usually allows for the best analysis and presentation product.

How Will It Be *Reproduced, Disseminated*, or *Viewed*?

There are three main considerations here: Will the map be viewed on a monitor, projected on a screen, or printed on paper? The map's mode of presentation especially affects the colors used, but also affects the layout and format. A map designed to be viewed on a monitor usually cannot be printed on paper without some loss of color fidelity—the colors look different. Solutions to this problem are discussed in Chapter 4.

For paper maps, one needs to know if the map will be printed by an inkjet printer, a laser printer, offset lithography, or some other method. There are differences in

FIGURE 2.9. Ortelius mapping software. Courtesy of MapDiva.com.

costs and time. If the map will be produced in large numbers, as with offset lithography, the cartographer should consult with a printer early in the design process.

Rules and Conventions

In designing maps there are a number of *conventions* and guidelines, but few rules. Conventions are such practices as blue for water, red for hot, and blue for cold. For some of these conventions there are logical reasons. Using red for hot, for example, is based on the idea that reds, oranges, and yellows are *warm colors* and blue and green are *cool colors* (see Chapter 4). Other conventions are based on old practices and have been used for centuries. For example, using red for urban areas supposedly originated in areas where building roofs were made of red tile.

Conventions are not rules and can be ignored, but only for good reasons. To use blue for hot and red for cold invites confusion, and coloring the oceans orange will draw the ire of most map users. On the other hand, showing a polluted river as brown would be a reasonable "violation" of the blue-water convention.

Intellectual and Visual Hierarchy

Not everything on a map is of equal importance. In the planning aspect of design one establishes an *intellectual hierarchy*. This is governed by the purpose of the map and its function.

If all elements are given equal visual weight the map becomes hard to read; it lacks contrast. As we have seen, maps are not linear and are not read in the way text is, from top to bottom and left to right. Establishing a *visual hierarchy* through size, boldness, and color helps lead the reader's eye (Figure 2.10). Thus, the mapmaker uses large type to attract the eye to the title and uses "heavy" colors such as red or black to emphasize areas.

One important aspect of the visual hierarchy is the *figure–ground relationship.* In a graphic communication, one area will stand out as the figure and another will be the ground or background. If the figure–ground relationship is not clear-cut, the communication will be ambiguous; this is the basis of many optical illusions (see Figure 2.11). If there is no clear visual hierarchy of color, an unclear figure–ground distinction can also result. For maps, the thematic information and the subject area are normally the figure and the base information is the ground.

The distinction between land and water is a special aspect of the figure–ground problem. Usually, the land is intended to stand out as figure, but if land and water have equal visual emphasis, readers have a very difficult time orienting themselves. Figure 2.12 illustrates coastal cities. Because the lettering is on both land and water areas, it is impossible to determine, without being familiar with the area, what is land and what is water.

Figure 2.13 shows several ways to establish a land–water distinction. *Water lining* was a customary way of symbolizing water on engraved maps for several centuries, but it is no longer an acceptable method unless one is attempting to create an antique feel. It is hard to read, and the lines are often mistaken for depth contours when, in fact, they have no numerical value.

Stippling is another conventional technique that was used primarily for manually drawn black-and-white maps. It was easy to do, and attractive when well done, but there is danger that readers might interpret the dots as representing sandy areas.

Line and *wave patterns* have also been used, but lines often create an unpleasant, vibrating effect that is hard on readers' eyes. The wave pattern is not desirable except

FIGURE 2.10. Visual hierarchy is established through size, boldness, and color.

FIGURE 2.11. A well-known optical illusion. Is the figure a vase or two profiles?

in very rare cases, such as cartoon maps or some pictorial maps. In addition to being hard on the eyes, wave patterns are considered childish and trite.

Color or *tone* are the best choices to distinguish land and water. Blue for water features is the most common convention, although even ocher has been used. Black oceans have been used effectively on maps, but there is risk that the water areas are given too much prominence and stand out as the figure when this is done. A gray tone applied to water areas is usually effective on black-and-white maps. *Drop shadows* appear to raise the land area and thus distinguish land from water or figure from ground.

The Search for Solutions

Creating a map is solving a spatial problem. How do I show these data most effectively or how do I tell this story? In fact, there are usually a number of different solu-

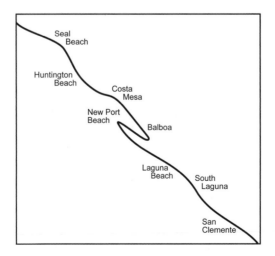

FIGURE 2.12. The land–water distinction is unclear on this figure.

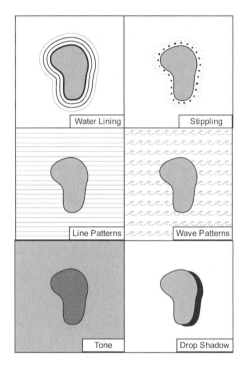

FIGURE 2.13. Ways of distinguishing between figure and ground (land–water).

tions that will work. In some cases, using two or more maps will be effective. For the Internet, one might create a multimedia or animated presentation. There is no single "best" map. If a single correct way to make a map existed, the topographic maps of all countries would look alike. An examination of these maps shows that while they share some features, there is a vast difference between Swiss, American, Dutch, German, and Mexican topographic maps, for example. Each country itself has determined what is most suitable for the maps of its area. While some might argue that one is more attractive than another, it doesn't hold that a "less attractive" map is wrong, poorly designed, or unsuited to the task. Swiss topographic maps represent the mountainous terrain of Switzerland beautifully, but the same techniques would not work on the flat topography of the Netherlands.

Once the cartographic problem is identified and understood, the search for solutions can begin. Preliminary "thumbnail" sketches can be of great help, even when making maps with the aid of an illustration or GIS program. These sketches help to create a graphic outline for the map. In the earliest stage, they may appear to be nothing more than doodles, but as the plan takes shape, these doodles can be expanded to form the layout of the map. Such sketches are not a waste of time; they are visual thinking (Figure 2.14). Computers, of course, allow quick tryout of solutions since elements can be moved easily.

Decisions are made at this time not only about the positions of the various elements, but also about the kinds of symbols to be used, color, map scale, and style and size of type. Decision making does not end here. At each stage of the mapping process it is worthwhile to analyze the design and fine-tune it if necessary to ensure that all elements are working harmoniously.

Design Constraints

Mapmakers do not have the freedom of design that other graphic designers do. The first constraint is the *shape of the area* represented. The shape of the United States cannot be altered to make it fit a given format. The area must remain recognizable. Different projections and orientations on the page provide some flexibility, but the projection used must still be appropriate for the map purpose.

Format and *scale* are also a constraint. Mapmakers are required to design maps to fit a specific format. A map that doesn't will be rejected by the editor. The mapmaker may also be required to make a map at a particular scale and this will govern how much area can be covered.

The amount of *text* required is also a restriction in map design. Some feel that maps would look much better without lettering, but place-names, legends, and explanatory text are usually necessary to clarify and identify features.

EXECUTION OF DESIGN (COMPOSITION)

Basic Elements

Once decisions have been made about projection, symbolization, and the like, the composition or layout of map elements can begin. The basic elements the mapmaker has to work with are the subject area, the title, the legend, the scale indicator, the graticule or north arrow, supplementary text, frame/border, and insets (Figure 2.15). Not all of these elements will appear on every map.

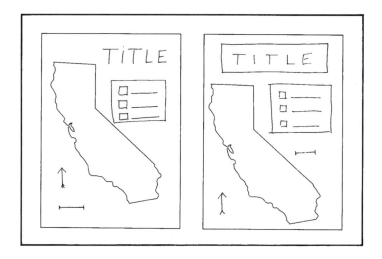

FIGURE 2.14. Thumbnail sketches are visual thinking.

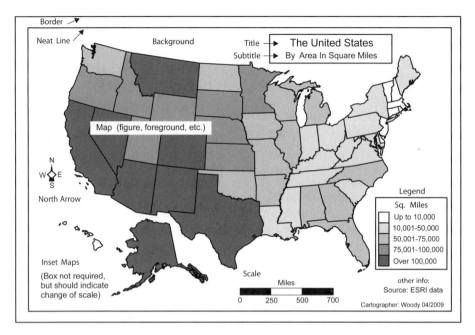

FIGURE 2.15. Generic map with the elements that can be used to make up a map design.

Subject Area

The subject area is normally the primary element of the visual hierarchy, it is the most important element on the page, and it is placed in the visual center of the page. It should also take up the most space within the frame. Often its place in the intellectual hierarchy is emphasized with graphic techniques such as drop shadows to raise the subject area above its surroundings, as in Figure 2.13. See also the figure–ground relationship above. The map should also provide a "sense of place" for the area.

Title

Most maps have a title. If the map is to stand alone, that is, printed on a separate sheet, not in a book, a title should appear on the map sheet; if the map is printed in a book, report, thesis, or dissertation, the title may appear on the map or as a caption below the illustration. The caption can explain or elaborate if there is a title on the map.

There are three things to consider with titles: wording, placement, and type style. The *wording* introduces the reader to the map subject just as the title of a book or article does. Wording and type style are covered in Chapter 3. *Placement* of the title is a part of the map layout. Contrary to what many believe, the title does not have to be at the top of the map. It can be placed anywhere on the page as long as it stands out in the visual hierarchy—the title is normally the most important wording on the map—and as long as it creates a balanced composition. The shape of the map area often provides a natural place for the title in the composition (Figure 2.16).

Legends

Legends present minidesign problems. Like title design, legend design has several parts: content, wording, placement, and style. First of all, any symbol in the legend must look *exactly* like the symbol on the map. Miniaturizing the symbol, for example, will cause reader confusion (Figure 2.17). It isn't necessary to title the legend space as "legend," although this is commonly done, especially on maps in children's textbooks, and was built into some early computer mapping programs. This is much like saying "a map of" in the title; it is redundant and a waste of space—although on children's maps it can serve as a teaching aid. The legend title can elaborate on the subject of the map and should explain the material in the legend (Figure 2.18). For example, if a map shows median income in the United States, by state, the legend could be titled "Income in Dollars." Or if the map title is simply "Income by State" the legend title can be "Median Income in Dollars." The goal is clarity (see Chapter 3).

Placement of the legend, like the other design elements, is governed by balance and white space. There is no general rule for where a legend should be placed although some companies and agencies may establish their own guidelines for a map series.

The lettering style of a legend does not have to be the same as that of the title, but the typefaces must complement one another. Some typefaces do not work well together (see Chapter 3).

Scale

In this section we are concerned with design and placement, not calculation and choice of scale; those topics are treated in Chapter 5. Scale can be expressed graphi-

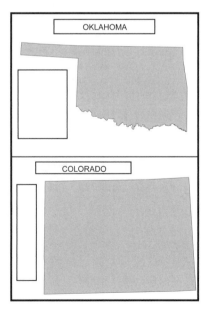

FIGURE 2.16. Some areas are easier to work with in design. Oklahoma provides a natural place for title and legend, Colorado does not.

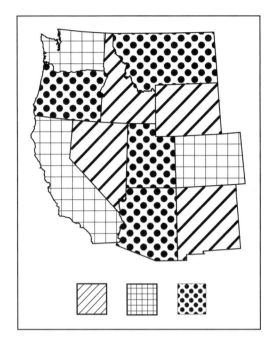

FIGURE 2.17. Symbols must look the same in the legend and the text or the map is confusing for the reader.

cally; as a bar or linear scale; as a verbal statement such as "1 inch represents 1 mile"; or as a fraction, such as 1:62,500. It is the graphic scale that most often causes design problems. First, one must remember that the scale is an aid to the reader, not the focus of the map. The scale serves one of two purposes: dimensionality or measurement. On a world thematic map, the scale indicates general size because the reader does not need to make precise measurements; on a large-scale map the reader might want to

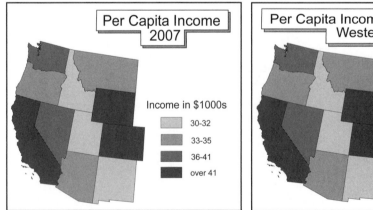
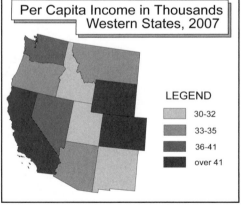

FIGURE 2.18. The word "legend" serves no purpose on the map on the right and it is better to replace it with a descriptive title.

know exact distances. Some computer software has a default scale that overwhelms the map and is the first thing the reader notices (Figure 2.19). The scale should be long enough to make necessary readings, but a 4-inch-long scale on a 6-inch map is overkill. Second, the scale should not be ornate. Ornate scales embellished with dividers were popular on 18th-century maps, but these maps also contained pictures of mermaids, sea serpents, and ships. Unless you are trying to imitate the feel of an old map, all such embellishments should be avoided. Figure 2.20 shows examples of acceptable scales.

The scale may be included in the legend area or it may be separate. As with title and legend, the scale is placed for balance and clarity.

Orientation

Orientation refers to showing direction, most commonly done by drawing the graticule (lines of longitude and latitude) or with a north arrow. Although it is a common custom, north does not have to be at the top of the map, and, in fact, sometimes cannot be. North has not always been at the top. European *mappae mundi* (world maps) placed the Orient (east) at the top and hence we have the term "to orient" the map. Early Chinese maps placed south at the top. The guideline now is that if there is no other indication, such as the graticule, north is assumed to be at the top and if it is not there must be some indication of orientation.

North arrows are a quick and easy way of indicating direction, but they must be used with caution. North arrows are not appropriate on all maps. For a small area like a city or neighborhood, they can be useful aids, but for maps of the world or large regions they may not be suitable. If the meridians on a map (true north–south lines) are curved or radiate, the north arrow is only correct for one point or along

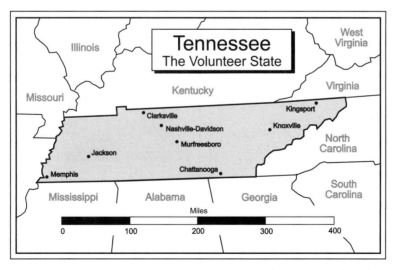

FIGURE 2.19. Oversize scales are common on many maps because of default options in the software. The scale is normally an aid, not the main focus of the map, and should not be a major element in the visual hierarchy.

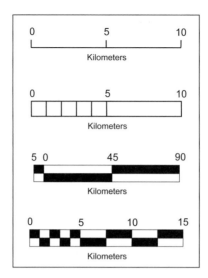

FIGURE 2.20. Graphic scales can be drawn in a variety of ways.

one line; on the conic projection shown in Figure 2.21, the arrow cannot be used. Unfortunately, improper use of a north arrow is a common error. Compass roses that show the cardinal directions (north, south, east, west) are generally used for navigation maps and are not usually appropriate for thematic maps.

If used, north arrows, like scales, are aids to the reader and shouldn't dominate the map. Many companies or agencies use a small logo for the arrow center and this can be effective, but still shouldn't overpower the map. Figure 2.22 shows a variety of north arrows and compass roses.

If parallels and meridians are drawn on the map a north arrow is redundant. The

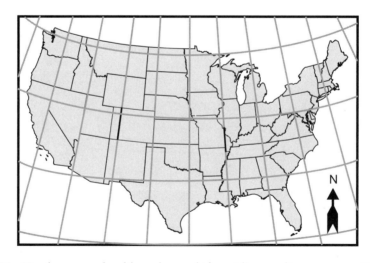

FIGURE 2.21. North arrows should not be used if meridians radiate or curve. The arrow on this conic projection does not point north.

FIGURE 2.22. North arrows and compass roses can take a variety of forms; usually simple forms are best.

choice of number of parallels and meridians depends on the scale of the map and its purpose. A map for navigation will require a finer grid than a general atlas map. On an atlas map the graticule serves to help the reader locate places. On a thematic map, grid ticks can be used because the reader isn't trying to determine precise locations. If grid ticks are used, they must be shown on all sides of the map because this also gives a sense of the type of projection (Figure 2.23).

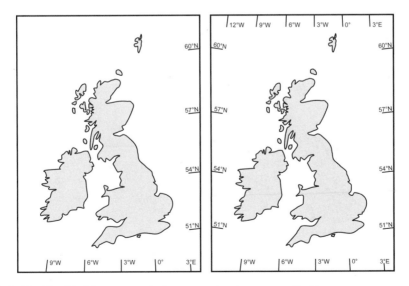

FIGURE 2.23. If grid ticks are used, they must be shown on all sides.

Inset Maps

An inset map is a small map used in conjunction with the main map and within the frame of the main map. Insets may be used to clarify, to gain scale, to enlarge or focus on a small section of the map, or to provide a setting for an area presumed to be unfamiliar to the reader. Inset maps can be quite helpful in solving difficult design and layout problems, but should not be overused (see Figure 2.24). Too many insets create a choppy, cluttered appearance and the design will not appear unified.

Some areas have irregular shapes that don't fit easily into a given format. Alaska is one such place. If the entire state of Alaska including the Aleutians is placed on a page in "portrait" format, the map is very tiny and it is difficult to show data; if an inset of the Aleutians is used, then the map can be larger. If the inset map is at a different scale than the main map, scales must be placed on both map and inset to avoid confusion about size.

Inset maps of Alaska and Hawaii are frequently used on maps showing the 50 United States. This sometimes creates a problem for children, who come to believe that Alaska and Hawaii are located to the south of the 48 contiguous states and that Alaska is an island. This problem can be solved by using an inset of North America and the Pacific with those two states highlighted. This type of inset is also used for any area that might be considered unfamiliar to the viewer (Figure 2.25).

In Figure 2.26 the detail of the center area cannot be distinguished; if the map is made large enough to show the detail area, it will no longer fit the format; if only the circled area is shown, the reader has no anchor for orientation. One solution is to enlarge the area and present it as an inset.

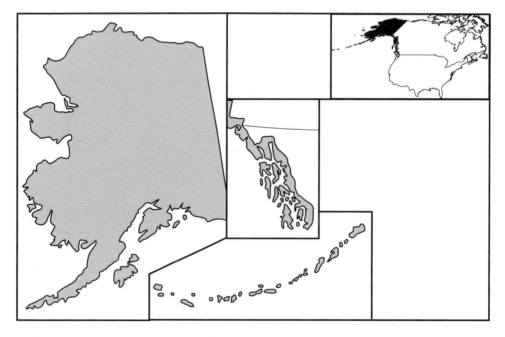

FIGURE 2.24. Insets are useful and information can be set off in boxes, but too many insets and boxes create a choppy, cluttered appearance.

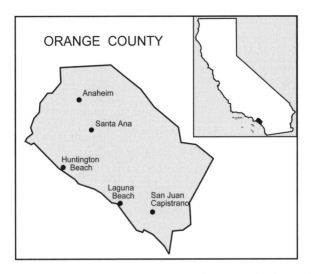

FIGURE 2.25. General location inset. The inset is used to provide the reader with the broad setting for an unfamiliar area.

Supplemental Text and Illustrations

There are elements of supplemental text that must be on the map. One example is a source statement, especially for quantitative maps or maps based on another's work. This statement acts in much the same way as a footnote in a book. Ethics require that such statements be included. Another piece of supplemental text is the name of the projection; this is an aid to the reader and provides a key to where the map is most accurate and what its limitations are. As with other elements, they are placed to create a pleasing well-balanced composition.

With the ease of manipulating photographs and text by computer, it has become increasingly popular to add large blocks of text and photographs to maps and atlases.

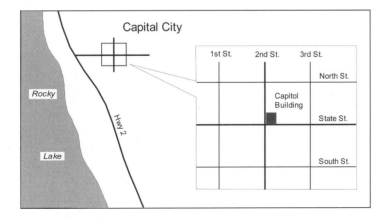

FIGURE 2.26. An inset can be used to focus on an area with an enlargement.

These can be effective, but should be used with care. The danger is including too much "stuff" on the page and losing the map in the process. Some modern atlases have pages containing more text than map. Viewers have mixed opinions on this practice. The key is to remember the focus and purpose of the map. Will the additional text add to the discourse or will it be visual clutter? Do the photographs illustrate the map subject or are they extraneous? There were other periods in cartographic history when decorative elements were included on maps. The seventeenth-century Blaeu atlases include drawings around each map page that showed dress styles for different cities or landmark buildings; other maps of the 17th and 18th centuries included allegorical figures. Nineteenth-century atlases frequently included vignettes showing historic events or local scenery. Figure 2.27 is a hand-drawn map with pen-and-ink sketches of buildings.

Frames and Neat Lines

The *neat line* is a line that frames and separates the map from other information; the *frame* is a border around the entire map. These lines are shown on the generic map in Figure 2.15. There is some debate about the use of frames and neat lines. Many like the "free" look a frameless map gives to the page, others feel that a frame adds stability to the page. To a large extent this is a personal preference, but is governed by

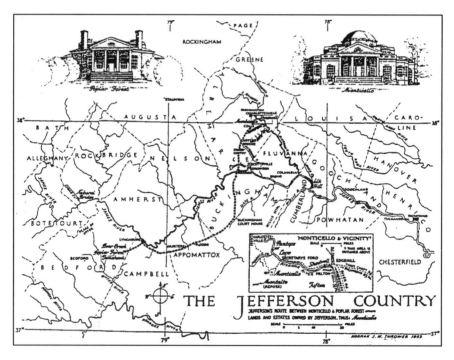

FIGURE 2.27. Drawings or photographs can add interest to a map if not overused. Hand-drawn map of part of Virginia with pen-and-ink drawings of Monticello. Courtesy of Norman J. W. Thrower.

where the map will be reproduced, who will use it, and the guidelines of the company or agency creating the map or map series.

Overall Appearance

Communication

The cartographer must consider the overall appearance of the map. It is easy to overload a map with so many symbols and topics that it becomes unreadable. It may be necessary to make more than one map to illustrate a given topic. On an overloaded map, the various themes and symbols fight for the reader's attention.

It is easy to overdesign graphics to the point where innovative technique or style dominates the graphics. A symbol on a map might be very eye-catching, but if the immediate reaction is to the appearance of a symbol instead of to its meaning, then there is a strong possibility that information has been lost. This is not to say that creativity and innovation should be avoided; an eye-catching map is more likely to be read than a humdrum one, but the goal of clarity must always be kept in mind.

Beauty

A significant aspect of the overall appearance that has often been overlooked or even considered unimportant in recent years is *beauty*. Old maps are often valued for their beauty and framed to hang on a wall, but it is hard to imagine an early computer map being treated in this way. Of course, beauty is hard to define and standards of beauty change, but readers still react to the aesthetics of maps.

We don't equate beauty in modern maps with ornate lettering, elaborate title cartouches, and drawings of mermaids and ships, but maps can still be attractive. Some maps are small, informal, and plain; they are devoid of ornament, because ornament would be inappropriate and distracting. These maps have elegance and beauty in their simplicity.

Other maps with a larger format, designed to show subjects of broad interest, such as many *National Geographic* special maps, benefit from the use of ornament. As we have seen, with restraint, ornament adds visual appeal, attractiveness, and pleasure—all worthwhile goals—as in good writing. Many maps are dull and sterile even if accurate. Good maps, like good writing, are enjoyable to view and satisfying to use.

Several organizations including the Cartography Specialty Group of the Association of American Geographers, the North American Cartographic Information Society, and Cartography and Geographic Information Science sponsor map design competitions each year. Consult their websites listed in Appendix B.

Spec Sheets

A useful habit (and standard practice with many companies and agencies) when creating a map is to keep a record of specifications: colors used, typefaces, line weights, statistical breakdowns. This can save hours of work when revisions or updates are needed.

Critiquing the Map

The last task in preparing a map is the critique. Even though you have checked every element at every step along the way, a final evaluation is in order.

SUGGESTIONS FOR FURTHER READING

Arnheim, Rudolf. (1969). *Art and Visual Perception: A Psychology of the Creative Eye.* Berkeley and Los Angeles: University of California Press.

Arnheim, Rudolf. (1971). *Visual Thinking.* Berkeley and Los Angeles: University of California Press.

Brewer, Cynthia A. (1994). Color Use Guidelines for Mapping and Visualization. In Alan M. MacEachren and D. R. Fraser Taylor (Eds.), *Visualization in Modern Cartography* (pp. 123–147). New York: Pergamon.

Brewer, Cynthia A. (2005). *Designing Better Maps: A Guide for GIS Users.* Redlands, CA: ESRI Press.

Brewer, Cynthia A. (2008). *Designed Maps: A Sourcebook for GIS Users.* Redlands, CA: ESRI Press.

Robinson, Arthur H. (1966). *The Look of Maps: An Examination of Cartographic Design.* Madison: University of Wisconsin Press.

Tufte, Edward R. (1983). *The Visual Display of Quantitative Information.* Cheshire, CT: Graphics Press.

Tufte, Edward R. (1990). *Envisioning Information.* Cheshire, CT: Graphics Press.

Chapter 3

Text Material and Typography

> . . . I am the voice of today, the herald of tomorrow. . . . I am the leaden army that conquers the world—I am TYPE.
>
> —FREDERIC WILLIAM GOUDY,
> *The Type Speaks* (1937)

PLANNING FOR LETTERING

Although maps are generally symbolic in nature, most maps contain considerable text material. Text on maps serves one of four purposes: (1) to label, (2) to explain, (3) to direct or point, or (4) to establish a hierarchy or show size—here type acts as a symbol. Therefore, the kinds of text materials we usually find on maps are titles, legends and explanatory material, statements of source, labels on symbols and scales, and place-names.

There are four aspects to lettering on maps: text wording, type placement, type selection, and editing. The text and lettering must be planned just like other elements of the design are planned. One cannot wait until the map is almost finished to decide what the title will be and where it will be placed; the space for the title must be allocated when the layout is first conceived. While computer methods make changes easier, it is better to take time in the planning stage to decide on wording and place-ment. If you fail to take time to plan, the result is crowded names, ill-conceived titles, and generally poorly executed design.

TEXT MATERIAL

Titles

Surprisingly, many published maps have inappropriate or even misleading titles. Although map titles are usually brief, they must relate the map topic clearly. The title is commonly the first element at which the reader looks. Unfortunately, many maps are produced for which the mapmaker seemingly had no understanding of the topic. Some time and thought are needed to develop an appropriate title that reflects the map purpose and yet is brief enough to fit into limited space. The map in Figure 3.1 clearly is intended to show the nature of the U.S. Public Land Survey System, but the cartographer has chosen to title it "Pecan Lake." The area shown is from the (mythical) Pecan Lake quadrangle, but the purpose of the map is not to show the lake, but to illustrate the numbering and lines of the Public Land Survey System.

Figure 3.2, titled "Oil Exploration," includes several symbols representing archeological sites. Either these symbols must be removed because they are irrelevant and a source of visual noise, or the map needs a title that more clearly characterizes the map purpose. If the map is designed to illustrate the threat posed to archeological sites by oil exploration, then the sites remain, and the map must be retitled to show that the relationship between the two features is the subject.

There may be three parts to the title: main title, subtitle, and perhaps a date. If both titles and subtitles are needed, as is common for series maps, which must stand alone or will appear in a book, the cartographer must decide what information should be emphasized, that is, the visual hierarchy of the text. Let us take as an example a population map of Nevada (Figure 3.3). Both "Nevada" and "Population" are needed in the title. Which should be given greatest emphasis, or should both be equally emphasized? If the map is to be one of a series of state population maps in a book on population, "Nevada" would be emphasized to distinguish it from other states. If the map is to be in a geography of Nevada, then "Population" should be emphasized. In the latter case, placing the name Nevada on each map would be redundant, so it

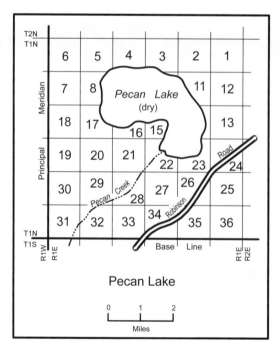

FIGURE 3.1. Map has an improper title. The map is designed to show a sample of the U.S. Public Land Survey System, not Pecan Lake.

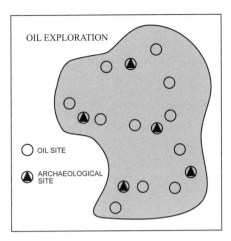

FIGURE 3.2. The title of this map doesn't reflect the subject of the map as shown by the symbols.

should be omitted. If a date is also needed in the title block, then its place in the hierarchy must be established. Abbreviations should be avoided, if possible.

The phrase "A Map of . . . " in a title, although frequently seen on old maps, is redundant. It should be obvious to the reader that a map is a map. Aside from being intellectually insulting, the phrase takes up valuable space.

Legends

The map legend contains explanatory text. As in the title, text in the legend should be concise, but it must also clearly explain the map symbols and their meanings. Many map readers are not experienced in interpreting certain kinds of symbols, so it is better to use a little extra space to clarify the map than to have the reader misuse or misread the symbols. Although the legend frequently has a brief title, as we saw in the previous chapter, the word "legend" or "key" serves no purpose and is redundant. It would be far better to elaborate on the map title and clarify the legend (see Figure 2.18).

FIGURE 3.3. Each of these titles can be appropriate depending on the subject and intent of the map.

Source and Other Explanatory Materials

Most thematic maps need a statement identifying the source of information. This is roughly the equivalent of footnotes in a written paper. A source for statistical data is needed so that the user who needs more precise tabular data can find it and also to indicate the authenticity of the data. Permission must be obtained for any copyrighted material, and a statement identifying the source and acknowledging permission to use it must be included on the map. Many publishers specify the wording of the permission statement—this wording must be followed exactly.

For world maps or maps of large areas, the name of the projection should be supplied. This tells the reader where the map is distorted and where it is most accurate. Without such information, it can be difficult to interpret a map.

LABEL PLACEMENT

Labeling includes the lettering attached to the scale, symbols, and place-names on the map. Graphic scales must have units attached or they are meaningless, but it is unnecessary to include the word "scale" or the phrase "scale in miles." This, like the words "a map of" and "legend," is redundant.

Guidelines for Name Placement

There are some conventions regarding the type of lettering used for some map features, and there are guidelines, although no hard-and-fast rules, for the placement of names and labels on maps. The main criterion for placement is clarity. GIS programs usually have functions that allow appropriate placement of features, but default options may not be the best choices. The following are generally accepted guidelines (Figure 3.4).

1. *Water features*, by convention, are labeled with *italic (slanted)* letters. *Rivers* are lettered with the name set in a block, not with spread-out letters, and the name is repeated at intervals, if necessary. If the river is represented by a single line, the letters should be placed, if possible, with the bottoms of the letters closest to the river. The direction of the slant of italic letters should be in the direction of the flow, if possible, while still permitting easy reading. The name of the river should follow the curve of the feature; therefore, for ease in placement, the name is usually placed in a straight stretch of the river, if possible.

Large *water bodies* such as lakes and seas should have the name placed totally within the feature. If the area is too small to do so, then the name should be totally outside the water body. The shoreline should not be broken. If curved parallels extend across an ocean, the lettering is most pleasing if it follows the curve of the parallel; otherwise, the appearance is very mechanical.

2. *Linear features* such as roads and railroads are labeled in much the same way as rivers, except the lettering is upright rather than slanted.

3. *Regional names* such as states and countries are normally spread out to

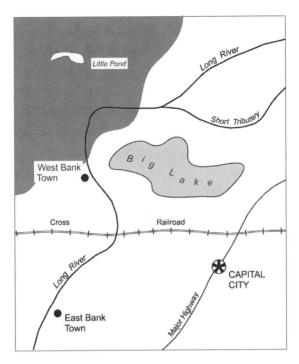

FIGURE 3.4. Preferred name placement.

cover the extent of the area and are curved in the shape of the area, as in Figure 3.5. The lettering is upright.

4. *Names of mountains* are treated in the same way as names of regions. In neither case is the name repeated at intervals as it is for linear features (Figure 3.6).

5. *Names of point locations*, such as cities, are placed to one side and not on the same line as the symbol. The first choice is above and to the right of the symbol. Generally, the name is not placed to the left of the symbol unless there is no other choice. The name is not placed on the same line as the symbol because there is a possibility that the symbol, especially an open circle, could be mistaken for a letter. These names are not spread out, but are set solid, that is, in a block (Figure 3.7).

If a city is located on the east side of a river, ideally the name should be placed on the east side, and places on the west side should have their names to the west. Obviously, this guideline can contradict the preferred locations above.

If it is not possible to place the name close to a point, a *callout* can be used, as in Figure 3.8.

6. *Lettering takes precedence over linework.* Lines should not be run through lettering, nor should names be broken to make room for lines. If there is a conflict, the line should be broken or a mask or halo can be placed around the letter (see below). On maps drawn in color, lettering can be put over linework if the lettering and the linework are different colors.

FIGURE 3.5. Regional names are spread to show extent.

7. *The primary rule is clarity.* These guidelines are not designed to be followed slavishly if the result is confusing.

Masks, Halos, and Callouts

Masks are rectangles that are placed under the type, but over the other graphics, creating special space for the lettering. These must be used with care because if they are too large they can obscure the underlying map information. *Callouts* are masks with *leader lines* that point to the feature. Although many shapes are available for callouts, simple ones are best. Unless there is a definite purpose, the cartoon balloon

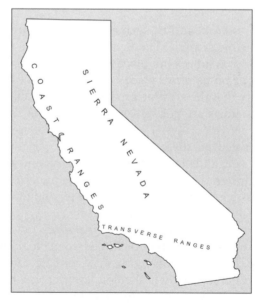

FIGURE 3.6. Mountain ranges are spread to show extent.

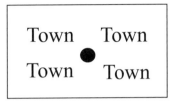

FIGURE 3.7. Names of point locations are not spread out and should not be on the same line as the symbol.

style should be avoided. *Halos* extend the outline of letters much like a drop shadow and make the lettering stand out (Figure 3.8).

GEOGRAPHIC NAMES

The spelling of foreign place-names on maps can be a challenge. There are two aspects to this problem: (1) transliterating names that are not in the Latin alphabet, such as Arabic, Chinese, Japanese, Greek, or Russian names, and (2) using names that, although in the Latin alphabet, are not the commonly used names in English, such as Wien for Vienna and Lisboa for Lisbon.

Non-Latin alphabet names may be transliterated according to guidelines specified by various agencies, but the correct local name and the name commonly used in the United States may be different. The local name may be unrecognizable to U.S. map users. If the map is to be used only in this country, there is normally no difficulty with using conventional anglicized names, but if the map is intended for international distribution, clearly the correct local name should be given priority, and the conventional name may be placed in parentheses. However, one must use caution here as international political incidents have been created by a map company or agency using the "wrong" name for a disputed place or territory.

A constant problem is that names are not static and may be changed for any number of reasons. New governments or even chambers of commerce alter names

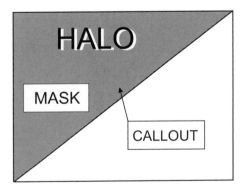

FIGURE 3.8. Callouts, masks, and halos can be useful when the background varies in tone or color.

to honor heroes, to reflect different ideals, to gain publicity, or to change an image. Thus, Ho Chi Minh City was formerly Saigon and St. Petersburg was formerly Leningrad. Many small towns in the Mother Lode area of California were originally given bawdy names by the miners, but as the towns became respectable, their names were changed. The town of Hot Springs in New Mexico changed its name in 1950 to Truth or Consequences because of a popular game show of the time.

A common error is including an English term that means the same as a part of the foreign name, such as Rio Grande River or Cape Cabo San Lucas. Table 3.1 lists some common foreign physical terms.

In the United States, the Board of Geographical Names is responsible for establishing guidelines on the spelling of names and will answer questions about names.

PRINCIPLES OF TYPOGRAPHY

Type styles are an important aspect of the appearance of a map and one frequently criticized by the user. "Hard to read" and "ugly lettering" are commonly heard complaints of map users. These criticisms can be avoided if the cartographer has a working knowledge of type styles and their psychological connotations and plans the map lettering carefully.

Terminology of Typography

As in many other fields, typography has its own vocabulary. Most of the terms describe the parts of the piece of type or type measurement. Although in map work actual pieces of type are no longer used, the terminology is derived from the early days of printing.

Type Measurement

The height of type is measured in points, with 72 points being approximately 1 inch. Thus one point is equal to 0.0138 inch or 0.39 millimeter. The size in points is not necessarily the height of the printed letter because the letter is generally somewhat smaller than the piece of type. Some typefaces are "large on the body," and others are "small on the body" meaning that they take up a larger or smaller portion of the original piece of type. Thus, two typefaces may both be 12 points, but letters on one may look (and be) larger.

TABLE 3.1. Common Physical Geography Terms

English	French	German	Spanish
Mountain	Mont, montagne	Berg	Montaña
River	Rivière	Fluss	Rio
Bay	Baie	Bucht	Bahia
Cape	Cap	Kap	Cabo
Lake	Lac	See	Lago

Parts of the Letter

Letters are positioned on a *base line* (Figure 3.9). The distance from the base line to the *mean line* is called the *x-height*, which is literally the height of the letter *x*; this measurement is important in determining type size for maps. Lowercase letters either rise to the *ascender line*, such as *b* and *d*, or extend down to the *descender line*, such as *p*, *q*, and *g*. *Serifs* are short extensions from the ends of the main strokes of a letter; and some letters have *bowls* and *loops* (Figure 3.10).

Fonts

A font is all the variations of a typeface of a given size possible. A well-developed typeface will include roman (upright) forms, italic (slant) forms, and differences in set or width, giving normal, condensed, and extended forms, and differences in weight, giving light, medium, and bold faces. Not all standard fonts available on computers include all of these forms (Figure 3.11).

Typefaces

Thousands of *typefaces* or styles are available, although not all are suitable for cartographic work. For ease of study, these faces may be placed into categories according to the appearance of the letters. A very simple categorization is serif or sans serif, but because there are some faces that don't fit into this categorization we can add decorative faces. These faces are only used on maps when a special effect is desired. Often these are advertising maps. Examples would be **comic book styles**, `Courier`, and *script styles*.

Another useful way of categorizing type is by the "mood" it presents, such as formal, informal, contemporary, or classic. Regardless of the names given to the categories, it is important to match the style of type to the style and purpose of the map.

Some authors describe as many as 11 categories of type styles, but for our purposes a more simple breakdown into four traditional categories will serve. These are *Oldstyle*, *Modern*, *Sans serif*, and *Special* types. The terms Oldstyle and Modern originally referred to type periods because the Oldstyle faces were developed in the 16th century and the Modern faces were first used in the 18th century. Now, new typefaces are placed into one of these categories according to general appearance, not date of design (Figure 3.12).

Oldstyle and Modern faces both have serifs and variations in line thickness. Oldstyle, however, has less difference between the thick and thin strokes and has a diagonal stress with asymmetrical bowls and loops. Modern faces are more geometric in appearance. There is a vertical stress and a strong difference between thick and

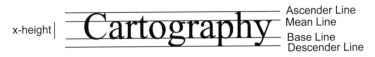

FIGURE 3.9. Letters are placed on a base line.

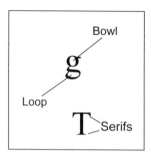

FIGURE 3.10. Parts of letters.

thin strokes; serifs are straight. Some writers include a transitional category between Oldstyle and Modern.

Sans serif faces, as the name implies, lack serifs. The letters may be all one thickness or there may be some variation of thick and thin strokes (Figure 3.13).

The special faces are grouped together here because they are only rarely used on maps. When they are used, it is usually to produce some special effect. In this category are found black letter styles, slab serif styles, and display faces (Figure 3.14).

Choosing Type

The major concerns when choosing a typeface are legibility, perceptibility, harmony, cost, convenience, and suitability for reproduction, and transferability or compatibility for online maps.

Type as a Symbol

The variations in size, form, spacing, orientation, value, and color of type makes it useful as a symbol in establishing visual and conceptual hierarchies and differences

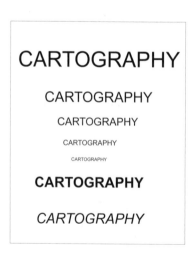

FIGURE 3.11. A font.

Old Style
Modern
Sans serif

FIGURE 3.12. Letter styles.

in kind. Size, tonal value, and spacing can indicate importance, while color and form indicate differences in kind. Orientation can be used to indicate direction. Larger type and bolder type indicate importance, form, for example, roman or italic, and hue indicate differences such as roads and rivers (Figure 3.15 and Plate 3.1). Textual hierarchy is also used for titles and subtitles (see also Chapter 7).

Legibility and Perceptibility

Most studies of legibility have been concerned with type set for text reading, not map reading. Perceptibility or discernability, that is, the speed at which letters or words can be perceived and recognized, is more relevant when choosing type for maps because the letters are not set in a block as they are for text. Instead, they may be letter-spaced (spread out), curved, or interrupted by other features. They may be displayed on shaded or colored backgrounds, and the background may even change in the middle of a word. When maps are viewed on a computer monitor the resolution of the monitor also affects the readability of the text. This presents a considerable challenge to the cartographer and, unfortunately, one that has no easy solutions.

Harmony

It is common to use more than one typeface on a map. However, usually only two faces, one serif and one sans serif (with the variations of bold and italic) work best and the styles used must harmonize with one another and with the subject of the map. It is generally felt that sans serif forms may be combined satisfactorily with serif forms, but that two serif forms usually do not work well with one another. Using too many styles creates a cluttered, messy appearance (Figure 3.16).

It is also accepted that different type styles have different personalities. Some are considered dignified, some are masculine, some are feminine, some are considered powerful, and some are weak. Obviously, this stylization can be carried to extremes, but one should at least keep in mind the possible connotations of a typeface because

Arial
Franklin Gothic
Lucida Sans

FIGURE 3.13. Sans serif typefaces.

Broadway
Old English Text
Calligraphy

FIGURE 3.14. Special typefaces are used for special effects.

the type can help set the tone of a map. Some descriptions of typeface personalities are shown in Table 3.2.

Typefaces also have periods of great popularity. For example, faces with great differences between thick and thin were popular in the 19th century. During the art deco period of the 1920s and 1930s, Broadway was popular. By using one of these faces, the map is given the feel of that period. Typefaces may also give a map a religious or antique look.

Although some recent studies have examined type selection and maps, cartographers must still rely on experience to a large degree. Looking at other maps is a great help in learning what works and what does not. A useful online tool in choosing typefaces is "Type Brewer," developed by Ben Sheesley, which allows you to see how your selection of typefaces work together on a sample map (Plate 3.2). A record of the various typefaces and sizes, that is, a spec sheet, should be maintained so that if changes must be made matching the type is simplified.

Balance and Hierarchy

Type has apparent weight, and words form lines. Dark, heavy type can act as a balance for other heavy areas on a map and draw the eye. Vertical letters act in the same way as a vertical line, and horizontal lettering leads the eye in a horizontal direction.

The various type styles can be used to attract the eye. The use of an unusual typeface will sometimes make the reader stop and take notice of an otherwise uninteresting map. Such ploys should be used only sparingly, however.

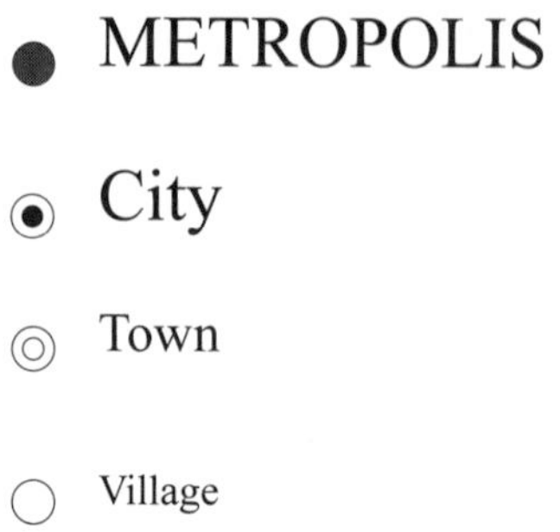

● **METROPOLIS**

◉ City

◎ Town

○ Village

FIGURE 3.15. Text can be used to show hierarchy.

Times New Roman

Microsoft Uighur

Goudy Old Style

ENGRAVERS MT

FIGURE 3.16. Many typefaces are available, but not all work well when used together.

Availability and Compatibility

There are thousands of typefaces, but not all will be available with any given software. Fonts may be purchased and added on to the software. There are also issues of compatibility or transferability with other computers if the map is done for the Web or on disk. If a particular font isn't available, the computer program will attempt to match it with one that is. This can result in a map that looks quite different from what you intended.

Suitability for Reproduction

Before choosing a type style, it is helpful to know how the paper map will be reproduced, if it will be reduced, and even, if possible, the type of paper that will be used. For maps that will be viewed on monitors or projected on screens, resolution is important. Generally, the Modern typefaces with extremes of thick and thin are considered less suitable for reduction than Oldstyle because the very thin lines may be lost in reproduction. They also do not work well on monitors.

EDITING

Like written work, the text of maps must be edited. The importance of editing cannot be stressed too strongly. One of the quickest ways to destroy the credibility of a map is to include misspelled words on it. The reader assumes that if the spelling is sloppy the research probably was also and that the map is likely to be unreliable. Therefore, all lettering on the map must be carefully examined. It is not enough to rely on a spell

TABLE 3.2. Type "Personalities"

Bodoni	Cold, austere, dignified, aloof
Goudy Oldstyle	Elegant, slightly feminine
Futura	Precise, graceful, aggressively modern
Scotch	Matter of fact, honest
Deepdene	Spicy and friendly
Centaur Italic	Sharp and dainty, clearly feminine

Note. Based on de Lopateki, Eugene. (1952). *Advertising Layout and Typography.* New York: Ronald Press. Hlasta, Stanley C. (1950). *Printing Types and How to Use Them.* Pittsburgh: Carnegie Institute of Technology.

checker in the computer program; it cannot determine if you have used the correct version of a name or if you have a correctly spelled but erroneous word.

The cartographer must also be sure that the sense of the text is correct, that is, that the words communicate the intended message. Therefore, after the spelling has been checked, the map as a whole must be examined for meaning.

Large mapmaking companies and agencies have quality assurance departments staffed with editors, but cartographers working alone or in small firms must perform their own quality checks.

SUGGESTIONS FOR FURTHER READING

Brewer, Cynthia A. (2005). *Designing Better Maps: A Guide for GIS Users*. Redlands, CA: ESRI Press.

International Paper Company. (2003). *Pocket Pal: The Handy Little Book of Graphic Arts Production* (19th ed.). Memphis, TN: Author.

Monmonier, Mark. (2006). *From Squaw Tit to Whorehouse Meadow: How Maps Name, Claim, and Inflame*. Chicago: University of Chicago Press.

Chapter 4

Color in Cartographic Design

"It's [color] got everything to do with it. Illinois is green, Indiana is pink. You show me any pink down there, if you can. No, sir; it's green."

—MARK TWAIN, *Tom Sawyer Abroad*
(1894)

Although color on maps has been seen as desirable from the early days of hand-colored maps, when maps were sold as "penny plain, twopence colored," until the 19th century color was not an option for mass-produced maps. Even into the 20th century color maps were costly and used only in large projects, such as atlases. Journals, magazines, and newspapers printed maps in black and white. Today, with the advent of computers, color monitors, and inexpensive color printers, color maps have become ubiquitous. Even the occasional mapmaker now needs to be familiar with the use of color on maps.

Color has a powerful visual impact; it attracts the eye and helps in directing the reader to the various elements of the map. Color allows greater flexibility in design; it aids in distinguishing figure–ground, such as land and water, and between categories, such as roads and railroads, rural and urban, and types of vegetation; it helps in establishing hierarchies within categories. While categories and hierarchies can be shown on black-and-white maps, the use of even one color on a map greatly enlarges design possibilities.

However, although the use of color is generally considered desirable, it does present the cartographer with some of the biggest map design challenges. Color, like typography, is one of the most frequently criticized aspects of a design. It is one of the most visible elements of a map. Readers tend to have definite likes and dislikes respecting color. Planning a color map is more complex than planning a black-and-white map; one must consider connotations, conventions, preferences, and interaction with other colors and with other map elements, such as typography, linework, and symbols. Registration of colors (alignment) can be a problem for printed maps, and preparation of the artwork is more complex.

THE NATURE OF COLOR

When we speak of "color" we are actually referring to the eye and brain reaction to a part of the *electromagnetic spectrum* (Figure 4.1). The electromagnetic spectrum is all energy that moves with the speed of light. It includes radio waves, x-rays, infrared waves, ultraviolet waves, microwaves, and what we refer to as the visible spectrum. Only a very small portion of the electromagnetic spectrum is visible to the human eye. This portion is called the *visible spectrum* and includes energy with wavelengths from 0.4 to 0.7 micrometers (μm). (A micrometer is one-millionth of a meter.) Light waves that have a length of 0.45 μm appear blue to us; those that have a length of 0.65 μm appear red. If a light source emits waves of all these lengths, the combination appears to have no color and is called *white light*. Light from the sun contains all the wavelengths. If white light is broken into its components, as with a prism or a raindrop, then we see a rainbow with the hues of the visible spectrum displayed in spectral order, from the shortest to longest wavelengths of violet, indigo, blue, green, yellow, orange, and red (Figure 4.2).

If light from single-hue sources of green, red, and blue are combined, the result is white light. Because these three wavelengths in various combination can produce any other hue of the visible spectrum or white, they are called *primaries*. Since the other hues are formed by adding the different wavelengths, they are called *additive primaries* (Figure 4.3 and Plate 4.1). The colors on a television screen or color monitor are formed by the additive primaries. These are used when working with maps on a computer screen. More research is being carried out in this area.

For conventional paper maps, we do not work with colored lights, but rather with inks or pigments on a sheet of paper. Mixing red, green, and blue pigments on a page does not result in white, but rather in a somewhat muddy hue that approaches black. When light strikes an opaque object, some of the light is absorbed by the object and some is reflected. A red object absorbs all but the red light waves, which are reflected. Pigment colors that can be combined on a page to produce all other hues are called *subtractive primaries* because some of the light waves are absorbed or subtracted by the paper. The subtractive primaries are magenta, cyan, and yellow (Figure 4.3 and Plate 4.1). These hues, plus black, are used in printing to produce any color or shade.

WAVELENGTH

GAMMA RAYS	X-RAYS	ULTRA-VIOLET	VISIBLE	NEAR INFRARED	SHORT INFRARED	MIDDLE INFRARED	THERMAL (RED) INFRARED	MICROWAVE	RADIOWAVES VHF TO LF

FIGURE 4.1. The electromagnetic spectrum.

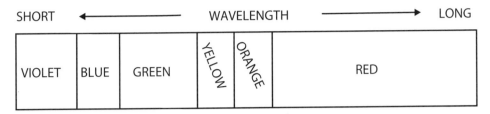

FIGURE 4.2. The visible spectrum.

DIMENSIONS OF COLOR

The term *hue* was used frequently in the previous discussion instead of *color*. Color actually has three dimensions, of which hue is only one. *Hue* refers to the sensation we perceive when light of a specific wavelength strikes the eye. Thus, red, green, blue, and violet are hues. But we recognize that there are many variations of red or blue. We commonly speak of "light blue" or "deep blue," "greenish blue" or "purplish blue." Some shades of blue seem somehow "bluer" than others. These descriptions represent attempts to describe the other two dimensions of color: *lightness* and *saturation*. Lightness is also called *value*, but when working with quantitative data that may have high or low numerical value the term can be confusing.

If we create a sequence of grays ranging from white to black with a series of progressively darker grays in between, we have produced a *gray scale* (Figure 4.4). The closer a gray is to white, the lighter it is or the higher its value; the closer to black, the darker it is and the lower its value. We can compare colors to the gray scale and establish lightness steps for different hues; a blue that is near the white end of the gray scale has a high value, and a blue that is near the black end has a low value (Plate 4.2). As we shall see in the next section, some color systems quantify the lightness of hues so that they can be compared and described.

Saturation, also called *purity, intensity,* or *chroma*, refers to the "colorfulness" of a hue. It is the extent to which the color deviates from a gray of the same value. A pure spectral color has a high intensity; it is fully saturated (Plate 4.3).

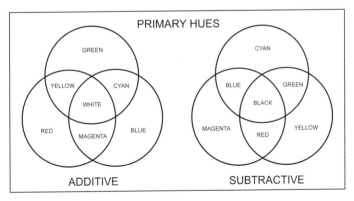

FIGURE 4.3. Additive and subtractive primaries.

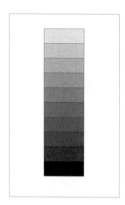

FIGURE 4.4. The gray scale.

Thus, when we speak of color, we are not only describing its basic hue, but also describing its lightness or darkness and its saturation.

COLOR SPECIFICATION SYSTEMS

In the previous sections we spoke of light or dark blue, greenish blue, or reddish blue and we identified colors as high or low lightness or high or low saturation. But if we attempt to describe a specific color, we are not sure our listener is visualizing the same color. Anyone who has attempted to describe a particular shade of blue to a painter and been surprised at the result is familiar with this problem. One solution, when dealing with a painter, is to provide a colored chip to match. But for many purposes, it is not feasible to provide color chips and a means must be found to describe colors. Several systems have been devised. I briefly describe three of the most common here, although anyone who intends to work intensively with color should consult one of the references for a fuller discussion.

CIE System

The Commision International de l'Eclarage (CIE), also called the International Commission on Illumination (ICI), utilizes a system based on standard illuminants, a standard observer, and standard primaries. Any color can be described precisely in numerical terms. It is used by the National Bureau of Standards and the National Geospatial-Intelligence Agency (NGA). It is based on the additive, not the subtractive, primaries; therefore, it is not as useful for printed map design as the other commonly used systems.

Munsell System

This system is commonly used in the United States. It was devised by Albert H. Munsell in 1898, and it has since been expanded and improved. In this system the three dimensions of color are utilized to create a *color solid*. Value forms the axis of the solid (Figure 4.5). Grays are arranged from white at the top to black at the bottom in a series of equal steps, which are numbered from 1 to 9. At the middle position on the gray scale (5) is middle or neutral gray; a plane passed through this point is a color circle with hues of middle lightness. Around the outer part of the solid, called the "equator" by Munsell, there is a decimal scale of hues merging gradually into one another, beginning and ending at red. The plane is also visualized as passing from the surface inward to the axis. The colors become less strong as one approaches the axis and stronger as the equator is neared. These represent the saturation, chroma, or strength of the color. Because not all colors have the same number of steps of saturation, the solid formed is not a sphere, but is asymmetrical. At each position on the gray scale, a series of rings shows different values. Each hue becomes lighter as one goes from the white pole of the solid and darker as one goes to the black pole. These rings are themselves color wheels. The hues arranged on the color wheel are specified by initial. Thus, any color can be specified by a letter and two numbers, and any new color can be assigned a place on the solid.

Spot Colors

While the systems above are useful for describing and understanding color, printed colors are chosen and described for the printer by a system analogous to the paint chips provided to a painter. Printers have a color specification chart that specifies the amounts of the three subtractive primaries and black necessary to create a specific color. The most common of these charts is the Pantone Matching System. Plate 4.4 shows the range of colors that can be produced from just two hues and shows the percentages of each hue used.

When creating colors at the computer and viewing them on screen, most illustration software programs have a function that gives quantities according to several different color models. CorelDraw allows CMYK (cyan, magenta, yellow, black),

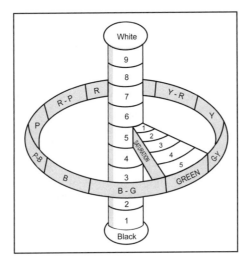

FIGURE 4.5. The Munsell system.

RGB (red, green, blue), HSB (hue, saturation, brightness), grayscale, and LAB (CIE); Adobe Illustrator CS allows HSB, CMYK, RGB, and grayscale. Other software has similar capabilities. Doing a print preview in grayscale lets the mapmaker see what the map would look like if printed in black and white.

CHOOSING COLORS

Color should not be used indiscriminately on maps. It should serve a purpose; if one invests the extra effort and cost called for by color, the effect should justify that extra effort and cost. Color should be anticipated in the stages of map design, not added as an afterthought.

Arthur Robinson (1967, pp. 50–61) gave three reasons for the use of color in cartography:

1. It acts as a clarifying and simplifying element. It increases the number of visual levels and acts as a unifying agent.

2. The use of color seems to have remarkable effects on the subjective reactions of the map reader.

3. It has a marked effect on the perceptibility on the map.

To these might be added:

4. It attracts attention.

5. It leads the eye.

Uses of Color

Color as a Clarifying and Simplifying Element

This is probably the reason most cartographers would cite for using color. Color is one of the visual variables of map symbols (see Chapter 7). Different values of a hue can be used to symbolize numerical values. Different-shaped symbols can be given the same hue to indicate objects within the same family. Hue can be used to distinguish symbols that have the same shape, such as different colored lines for different routes or kinds of transportation lines. Colored dots can be used to illustrate two different products or crops on a dot map.

Closely tied to this reason is the use of color to *enhance perceptibility* on a map. Figure 4.6 illustrates a black-and-white route map. Although one can distinguish the different line patterns and they are explained in the legend, the map would be much clearer if even one additional color had been used to differentiate the lines (Plate 4.5) Color is also one of the simplest ways to create a clear figure–ground relationship and establish visual hierarchies.

FIGURE 4.6. The use of color would enhance the perceptibility of this route map.

Attention Attraction

We really do not know why color attracts the eye and is an *attention* getter, but it is undisputed that this is so. It is also not clear which colors have the greatest attractive power; there are conflicting test results and opinions. Generally, it appears that the attention-getting value depends on visibility. Therefore, the color that can be seen at the greatest distance is the one that attracts the eye quickest. It has also been pointed out that pure hues have greater visibility than tints or tones. However, visibility also depends on the background. Yellow when next to a blue-black is highly visible, but when printed on a white page it has very low visibility. There is also a difference between the color used for areas, for typography, or for lines or points.

Study of actual printed maps reveals that cartographers, whether intellectually or instinctively, accept red as the color highest in attention value since this is the color most often used when only one color is chosen in addition to black and white. Many studies certainly indicate that red rates highly as an attention getter. Some textbooks have used brown or green in addition to black and white on maps. While brown is generally rated fairly low in attracting the eye, the mere fact that it is the only color on the page makes it stand out. When only one color is used on a map, such as a red arrow or one country or region in red, the purpose is usually to draw attention to a specific feature and to lead the eye.

Color Preferences

As mentioned above, color is one of the most frequently criticized aspects of a map. People have definite color preferences. However, although studies have been done testing color preferences, they often look at color in the abstract, not color used for specific purposes. Thus, a person might like red cars, but not want to sleep in a red bedroom. Preferences appear to change with age, different cultures seem to have different preferences, and men and women have different preferences. Although the literature on color preference dates back to work by Goethe, and supposed preferences have been exploited on propaganda maps (Tyner, 1974), color preference is normally not a good basis for choosing colors on maps.

Color Associations and Connotations

Color associations and color connotations can be considered, but must be used with care. Colors have been associated with a variety of attributes, such as smell and temperature, but many of these associations are either individual or cultural. Temperature is one of the few associations that approaches universal: colors are commonly described as "warm" or "cool." Reds, oranges, and yellows are warm colors and blues, purples, and greens are cool colors. This association is strong enough that maps showing temperature typically use warm colors for hot temperatures and cool colors for cold ones. Blue for water is another common association that is used almost universally on maps, even for polluted water features.

Sometimes these conventions create confusion. Climate maps commonly represent desert areas in shades of tan or orange and rainy areas in blue. Hypsometric (elevation) maps, which show only elevation, frequently represent lowlands with greens and highlands with reddish browns. This is based on the idea that cool colors appear to recede and look farther away and warm colors look closer to the viewer. Unskilled readers often interpret these hypsometric colors not as indications of elevation, but as representations of vegetation, climate, or rainfall. The green areas (lowlands) are perceived erroneously as moist and lush and the brown areas as hot and dry. Clear legends are important.

Colors also have connotations, such as red for danger and yellow for caution; we speak of "feeling blue" or "seeing red." These connotations can be taken into consideration and have been used on maps, but again, one must remember that these are not universally accepted meanings and vary by culture.

It might seem that associating red with courage and yellow with cowardice would have no relationship to the use of color on maps, but coloring a country yellow on a map might prompt an unfavorable reaction because the people of that country are offended by the supposed symbolism. It is always a serious mistake to use stereotypical skin colors for maps showing racial patterns because the likelihood of causing offense is great. It is easy to use more abstract colors.

In the United States, income and other financial subjects are often shown in shades of green. This makes some sense because U.S. paper money is primarily green, but it doesn't work in other countries where the currency might be pink or purple.

It is impossible to take into account all the possible reactions to specific colors, but it is helpful to be aware of some of the more common color associations. Henry

Dreyfuss's *A Symbol Sourcebook* is a valuable reference for the symbolic meaning of color.

As with other design decisions, when choosing colors for a map, ask who is my audience, what is the nature of the data, where and how will the map be viewed, and what is the purpose of the map?

Audience

When considering the audience, the user's visual abilities, color preferences, color associations, and color connotations come into play. Color blindness affects a portion of the population. There are varying kinds of color blindness, such as red–green, blue–yellow, and monocromatism. The percentage of color blindness within a society varies from culture to culture, but generally more men are afflicted than women. While normally one isn't designing a map strictly for that segment of the population, it is wise to choose colors that do not create problems.

Viewing the Map

Where and how the map will be viewed is a major consideration. As we have seen, primary colors differ for light and pigment and a color scheme that looks good on a high-quality monitor will look quite different when printed. Colors on a monitor, because they are based on additive primaries, look different from printed colors, which are based on subtractive primaries. Therefore, even if designing a print map on screen using a CMYK pallette, the color scheme needs to be tested on paper. Maps that will be viewed through a data projector, as for PowerPoint projections, need to be designed for that purpose (if possible). Unfortunately, the cartographer has no control over the type of projector used and how it has been calibrated, but some schemes work better than others. ColorBrewer (see below) suggests color schemes that are appropriate for projectors.

TABLE 4.1. One-Variable Data Types and Color Schemes

Data conceptualization and scheme type	Perceptual characteristics of scheme	
	Hue	Lightness
Qualitative	Hue steps, not ordered	Similar in lightness
Binary	Neutrals, one hue or one hue step	Single lightness step
Sequential	Neutrals, one hue or hue transition	Single sequence of lightness steps
Diverging	Two hues, one hue and neutrals, or two hue transitions	Two diverging sequences of lightness steps

Note. From Brewer, Cynthia A. (1994). Adapted by permission.

Nature of the Data

The nature of the data is a major factor. Data may be qualitative—that is, non-ordered, or quantitative, ordered and numerical—and within these categories there are subcategories. Table 4.1 is based on Cynthia Brewer's work with color schemes and shows four different data conceptualizations and their associated color schemes. ColorBrewer (*www.ColorBrewer.org*) is a useful site for learning how to understand and choose color (Plate 4.6).

Qualitative Schemes

Qualitative schemes are used for data that have no magnitude or size difference between classes; they are *unordered* and show only qualitative information (differences in kind), such as land use or vegetation types. Qualitative schemes show differences between categories with different hues; lightness should be similar, although small lightness differences are necessary for the colors to be distinguished from one another and allow areas to be identified more easily. There should not be large differences in lightness or saturation because these imply differences in importance.

Since the advent of GIS, which makes use of color easier, maps are frequently created that have as many as forty qualitative categories, often made up of varying lightness steps of hues. These maps are unreadable; the user tends to group the reds together, the greens, the yellows, and so forth, making perceptual categories. If there is no logic to the hue/lightness choices, a strange impression, indeed, is the result. An example of lack of color logic is shown in Plate 4.7 Red is used for coniferous trees, pink for grasslands, green for deciduous trees, light green for bare land, brown for mixed forest, and tan for shrubland. While this is an extreme example, many published maps make similar errors.

A qualitative map with many categories presents a major challenge and requires thought and experimentation to achieve a clear map. If possible, the number of categories should be reduced, but if this is not possible, then lightness and saturation differences can be introduced, but by using some logical system. For example, urban and suburban areas imply an ordered relationship, so that lightness steps of one hue, red, perhaps, can be used to distinguish the categories. For the vegetation example above, using green for all trees makes sense, with the variations of deciduous and coniferous shown with variations in saturation and lightness.

Brewer considers *binary schemes* to be a special case of qualitative schemes. Binary variables are those which have only two categories, such as rural/urban, private/public, and populated/unpopulated. For these categories she suggests one hue or two hues or a neutral with the difference between them a lightness step (Plate 4.8)

Quantitative Schemes

Quantitative maps are those that represent some numerical aspect of a spatial distribution, such as temperature, rainfall, elevation, and the like.

If data show a progression from low to high, a simple *sequential color scheme* is normally used. The scheme may be made up of neutrals, that is, shades of gray, one hue, such as shades of blue, or a transition of two hues, such as yellow to blue (Plate

4.9). In these cases the differences are of lightness. There is a limit to the number of lightness steps that can be perceived, so the number of categories is usually restricted, and the number possible depends upon the hue or hues selected. Most cartographers use between five and seven steps for quantitative maps (see Chapter 8). Yellow has fewer distinguishable lightness steps than red. A single hue permits the fewest distinguishable steps and a two-hue sequence permits the greatest number. Thus, if more than seven steps are needed, a two-hue sequence is a good choice.

Some data do not progress from high to low, but are seen as two-ended or diverging from a midpoint. Positive and negative change with zero as a midpoint is an obvious example of this. Other examples include data above or below a median, or average temperatures above or below freezing. For these data a *diverging scheme* works well. Diverging schemes basically utilize two sequential schemes joined at the critical or midrange figure. Some examples are shown on Plate 4.9.

Color Interactions

Colors are not viewed in isolation on a map. Therefore, one must consider the ways in which colors interact with one another on the map. One example is *simultaneous contrast*, which refers to the illusion that a color looks different based on the surrounding colors. A gray square will look lighter or darker depending upon the color that surrounds it (Plate 4.10). Thus, a color may appear different on the map than in its legend. The grays of the neutral scale in Figure 4.4 appear lighter when next to darker grays and darker when next to lighter grays. This problem can be partially solved by enclosing the areas with lines. Areas that are very small on a map require more lightness in color in order to be visible.

Lettering on a colored background can create a special problem (Plate 4.11). One must be especially cautious when using colored lettering on a colored background. Some combinations can be difficult to read and others are jarring to the eye. Black on yellow is considered the most legible combination and is frequently seen on road signs for that reason. Green or blue, especially dark green or blue, on white is next on the scale of legibility. At the bottom of the scale are red on green, green on red, and red on yellow. These combinations should be avoided because they are jarring to the eye. This also applies to stripes or other patterns. Red and green stripes are disturbing to the eye and alternating dark pink and red or dark and light yellow may be difficult to distinguish.

Black and White

As noted in Chapter 2, color is not always an option. Although color has become common for online mapping, and for a few copies produced on a printer, there are still many occasions when black-and-white maps are required. Even though journals, magazine, and newspaper maps are increasingly produced in color, it is still more expensive. The majority of maps in textbooks and other books are still created in black and white. Thus, it is necessary to be aware of some of the considerations for black-and-white design.

A major consideration is distinguishing different shades of gray. In Figure 4.7, the grays in the legend are easily distinguished, but they are not when on areas of the map

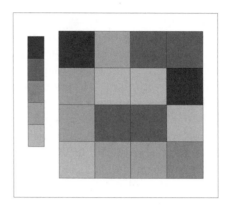

FIGURE 4.7. Grays can look different on a map and in the legend.

that are separate from one another. In Figure 4.8 the two squares are the same gray, but when they are placed on darker or lighter shades, they look different. In addition, although it is easy through software to create 10 shades of gray, the human eye cannot distinguish that many shades easily. The gray scale in Figure 4.4 shows 10 lightness steps, but adjoining steps are not easy to differentiate from one another. This is one reason that most quantitative black-and-white maps use only five or six shades.

Because lightness steps usually imply quantitative value, using different shades of gray for different qualitative categories is misleading. In such cases, patterns can be used and the orientation of lines can be varied (although this can create an "old-fashioned" look). Pictorial area symbols can also be used, such as the grassland and tree patterns in Figure 4.9. These visual variables are described in Chapter 7, on symbolization.

If one knows at the beginning of the design process that the map will be limited to black and white, it is easiest to simply design in those neutrals. However, if the map may be viewed in both black and white and color, then colors that reproduce in distinct shades of gray must be chosen. This requires some care. For example, if a color map might be printed in both color and black and white for different purposes, or if it is likely that black-and-white copies will be made on a copy machine, for example, handouts of a PowerPoint presentation, this must be considered before choosing colors. The ramp (shades of gray) may not be a smooth gradation or the

FIGURE 4.8. Grays look different depending on what is next to them. The inner squares are the same shade of gray.

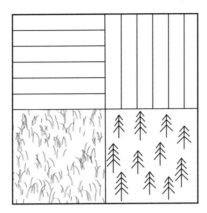

FIGURE 4.9. Symbols for a black and white map.

number of steps in the color sequence may need to be limited (Figure 4.10 and Plate 4.12). ColorBrewer can again assist with choosing colors that are copier-friendly or print-friendly.

Specification Sheets

A record should be kept of the colors used and their specifications. ColorBrewer provides printable "spec sheets" for various color schemes, but one can also design one's own sheet that includes typefaces and sizes and line weights.

Color Esthetics and Harmony

A final aspect to consider is the overall harmony of the map. A map with pleasing color combinations isn't merely more attractive, it's easier to read than one with

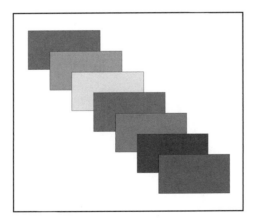

FIGURE 4.10. If a map is to be reproduced in color as well as black and white, colors must be chosen carefully; a color map cannot simply be printed or copied in black-and-white without loss of information. This figure is Plate 4.12 printed in black and white.

clashing colors. Check to see that all lettering is visible against its background, that small areas are easy to see, but not given undue importance, and that there is logic to the color choices.

SUGGESTIONS FOR FURTHER READING

Brewer, Cynthia A. (1994). Color Use Guidelines for Mapping and Visualization. In Alan M. MacEachren and D. R. Fraser Taylor (Eds.), *Visualization in Modern Cartography* (pp. 123–147). New York: Pergamon.

Brewer, Cynthia A. (2005). *Designing Better Maps: A Guide for GIS Users*. Redlands, CA: ESRI Press.

Brown, Allan, and Feringa, Wim. (2003). *Color Basics for GIS Users*. New York: Prentice Hall.

Dreyfuss, Henry. (1972). *A Symbol Sourcebook*. New York: McGraw-Hill.

International Paper Company. (2003). *Pocket Pal* (19th ed.). Memphis, TN: Author.

Linford, Chris. (2004). *The Complete Guide to Digital Color: Creative Use of Color in the Digital Arts*. New York: Harper Collins.

Robinson, Arthur. (1967). Psychological Aspects of Color in Cartography. *International Yearbook of Cartography*, 7, 50–61.

PART II

THE GEOGRAPHIC AND CARTOGRAPHIC FRAMEWORK

Chapter 5

Scale, Compilation, and Generalization

There's no escape from the cartographic paradox: to present a useful and truthful picture, an accurate map must tell white lies.

—MARK MONMONIER, *How to Lie with Maps* (1996)

SCALE: BRINGING THE EARTH DOWN TO SIZE

All maps are drawn to *scale*, which means they are drawn smaller than reality.[1] This is part of what makes maps useful; they reduce the earth to a manageable, comprehensible size. *Map scale* is the size of the map compared to the size of the real world. We use a fraction to describe the relationship, such as 1/1,000,000 or 1/24,000. These mean the map is one millionth as large as the real world or one 24 thousandth the size of the earth. When we create a map, the scale is a part of the geographic framework.

Scale can be described in three ways: as a ratio, in words, or graphically (Figure 5.1). A ratio is simply a comparison of map size to real-world size as in 1:1,000,000 or 1:24,000. This means one unit on the map represents 1,000,000 of the *same* units in the world; thus 1 inch represents 1,000,000 inches or 1 centimeter represents 1,000,000 centimeters. It is the same as saying the map is 1/1,000,000 the size of the earth, as above. Expressing scale in this way is called the *representative fraction* (RF), the natural scale, or *the* scale of the map. No units (inches, centimeters, miles, kilometers) are attached to the RF since it is a ratio.

Because numbers like 1/1,000,000 or 1 inch represents 24,000 inches are hard

---

[1]Drawings of objects seen in a microscope are also drawn to scale, but the scale is larger than reality. Maps are never drawn larger.

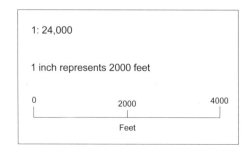

FIGURE 5.1. Scale can be expressed in three ways: as a ratio or fraction, verbally, or as a linear or bar scale.

to visualize, the scale is often converted to words. Thus, 1/1,000,000 becomes 1 inch represents about 16 miles or 1 centimeter represents 10 kilometers; 1/24,000 becomes 1 inch represents 2,000 feet or 1 centimeter represents 240 meters. This is called a *verbal scale* or a *word scale*, in which the figures may be rounded off. For example, 1:1,000,000 is actually 1 inch represents 15.782828 miles but would be rounded to either "about 15 3/4 miles" or "about 16 miles." Because verbal scales are often used as a rough estimate, such rounding doesn't usually present a problem.

A third way of expressing scale is visually or graphically. A line is drawn at the same scale as the map and divided into units. Thus, for a 1:24,000 map, a one-inch line represents 2,000 feet and 1/2 inch represents 1,000 feet. This kind of scale is called a *graphic scale, bar scale,* or *linear scale.*

The terms "large scale" and "small scale" are often used and frequently confused. A large-scale map shows a small area in great detail; a small-scale map shows a large area but with little detail. Thus, a world map shown on an atlas page is a small-scale map and a city map appearing in the same atlas is a large-scale map. The terms large scale and small scale are relative terms; there is no specific dividing point for large and small scales, although some map series such as USGS topographic maps will have set scales for large, medium, and small scales *for the series.* The larger the denominator of the RF, the smaller the scale. The scale 1:1,000,000 is smaller than the scale 1:500,000. This is the same as any fraction: 1/6 of a pie is smaller than 1/4 of a pie. Figures 5.2a, 5.2b, and 5.2c show the same general area at three different scales. Note the size of the area shown, and the amount of detail.

Calculating Scale

Often the scale of a map must be calculated. If a base map is available that has only an RF and the cartographer wants to know the verbal scale or to draw a linear scale or if the scale of a map isn't known, the scale can be determined. Table 5.1 has some useful conversion factors.

Examples

Determine a verbal scale for a map with an RF of 1:253,440

1 inch represents 253,440 inches.

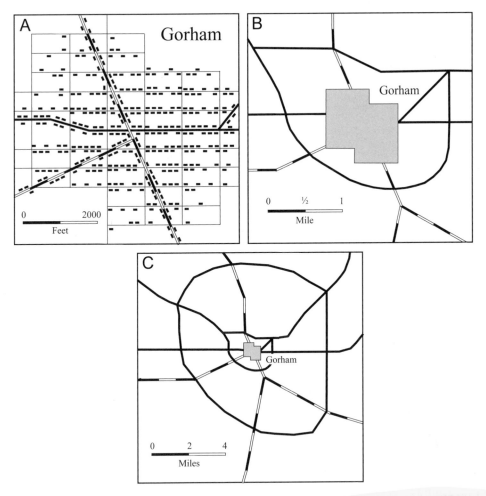

FIGURE 5.2. The maps in this figure are all of the same area, but at three different scales. Note the amount of detail that can be shown at different scales.

To find the verbal scale in miles, one divides 253,440 by the number of inches in one mile

253,440 ÷ 63,360 = 4. Therefore, 1:253,440 is the same as one inch represents 4 miles.

To find the verbal scale in kilometers for the same map, one divides 253,440 by the number of centimeters in 1 kilometer.

253,440 ÷ 100,000 = 2.53. Therefore, 1:253,440 is the same as 1 centimeter represents 2.53 kilometers.

Another kind of problem occurs when one has a source map or photograph that has no scale. Here, two known points of known distance must be found. These can be two points on a border, two roads, or, on photographs, some feature with known

TABLE 5.1. Conversion Factors

1 international nautical mile	=	6,076.1 feet
1 nautical mile	=	1.15 statute miles
1 nautical mile	=	1.852 kilometers
1 statute mile	=	5,280 feet
1 statute mile	=	63,360 inches
1 statute mile	=	1.61 kilometers
1 statute mile	=	0.87 nautical miles
1 kilometer	=	0.54 nautical mile
1 kilometer	=	0.62 statute mile
1 kilometer	=	100,000 centimeters
1 kilometer	=	1,000 meters

dimensions, such as a football field. The map distance between the points is measured and using the formula RF = Map Distance/Ground Distance the scale can be determined. Remember that RF is a fraction expressed as 1/x. This method works best for large-scale maps, and is only approximate for large regions depending on the projection.

Representing Scale

On a map, scale may be shown in any or all of these three ways. Topographic maps and atlases frequently include all three forms of scale. Moreover, the graphic scale may be shown in both English and metric units. However, on thematic maps and in reports, periodicals, and newspapers, it is most common to use only a graphic scale. Why? Because the cartographer isn't always sure at what size the map will be reproduced—Will it be full-page, one column, or 1/2 page?—or even what size the page will be. If the map is printed larger or smaller than its original size, the RF and verbal scales will no longer be accurate. However, a graphic scale will enlarge or reduce in the same proportion as the rest of the map. *For maps that will be viewed on a monitor, the only appropriate scale is a graphic scale.* Monitors vary in size; therefore, the map will be larger or smaller depending on the size of the screen.

A common mistake many novice mapmakers make is creating a graphic scale that is more precise than the map (Figure 5.3). If a thematic map is very generalized and at a small scale, such as 1:1,000,000, there is no advantage to making a bar scale marked in 1-mile units. Measurements this small cannot be made from the map and it gives an incorrect impression of the map's accuracy. Even at a scale of 1:24,000 (1 inch represents 2,000 feet) a 20-foot-wide stream drawn to scale would be .01 inches wide and would normally be drawn wider for legibility.

Choosing Scale

Several factors impact choosing a scale for a map: the subject and purpose of the map, data resolution, map user needs, and the specified format.

If a map is designed for navigation or hiking or bicycling, it will need more detail than if it is designed to show an overview of a water body, a national park, or a recreation area. Bill Bryson in *A Walk in the Woods* expressed his frustration with maps that were at too small a scale for the intended use:

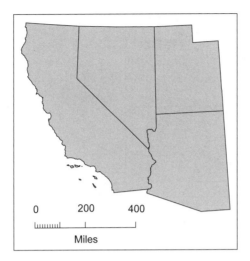

FIGURE 5.3. The graphic scale should not be more precise than the map. Detail on this map cannot be read at 10-mile increments.

> I dumped my pack and searched through it for my trail map. . . . [The maps] vary some-what, but most are on an abysmal scale of 1:100,000, which ludicrously compresses every kilometer of real world into a mere centimeter of map. Imagine a square kilometer of physical landscape and all that it might contain—logging roads, streams, a mountain-top or two, perhaps a fire tower, a knob or grassy bald, the wandering AT [Appalachian Trail], and maybe a pair of important side trails—and imagine trying to convey all that information on an area the size of the nail on your little finger. (pp. 73–74)

Bryson's frustration clearly shows the necessity of considering user needs when choosing scale. To Bryson, the hiker, the details of the trail he was following were vital to his survival and a scale of 1:100,000 wasn't sufficient. On the other hand, for a map user who wants to get a broad overview of the length of the Appalachian Trail, the states it covers, and its general location, a scale of 1:100,000 is far too large.

The resolution or detail of the data represented also is of importance in choosing a map scale. If a great deal of detail must be represented, then the scale ideally will be larger than if the data resolution is less.

At times the page format is specified and the map must fit within its confines. If the area is large, the cartographer has few options other than including larger scale insets to show the details of some areas.

A common problem is using a map that was designed for a different scale. This problem frequently occurs in newspapers, periodicals, and websites when a map published by an agency or individual for use at one size is reduced to fit the column inches available (see Figure 5.4). Much information is lost through this reduction. Ideally, the map should be redrawn to fit the format, but time constraints may dictate that the original be used. In such cases, the source should be cited so the reader can go to the original.

As we will see, map scale is an important factor in choosing the amount of generalization that will be used on a map.

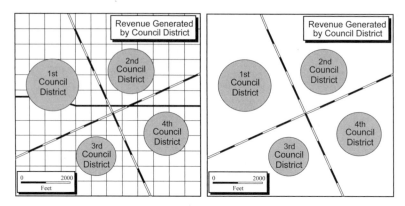

FIGURE 5.4. If a map is to be printed at a reduced scale some information must be removed.

COMPILATION: GATHERING THE DATA

Compilation is collecting and selecting all essential information needed for the preparation of a map. There are two stages in the compilation process: deciding what data are needed and gathering those data from a variety of sources.

Kinds of Compilation Data

Two kinds of information must be collected when creating a map. These are base data and thematic or primary data.

Base Data

The base data present a background reference for the thematic information in order to make the theme of the map clearer and more easily understood. Base data are not the focus of the map. They provide a structural framework for the design and help the map reader to interpret and elaborate the purpose of the map. It is important that only information that will help the reader be included in the base, but that enough background data is supplied. Too much or extraneous information serves only to clutter the map and is a source of confusion. Too little base information frustrates the reader, making it difficult to relate the thematic information to the real world.

Dennis Fitzsimons has made a useful classification of base data as "internal" or "external." *Internal base data* include such things as administrative boundaries, coastlines, cities, transportation routes, place-names, mountains, rivers, and lakes, that is, geographic information on or within the mapped area itself. For a series of maps these data form the base layer that will be used for all maps in the series and together are called the *base map. External base data* include title, legend, scale, north arrow, grid, and text. The external base data have an explanatory function. These have been discussed in Chapter 2 as design elements and are also discussed in other chapters. Therefore, only internal base data is discussed here. The projection chosen is an important part of the base map; this topic is discussed in Chapter 6.

The decision regarding how much base data and what type to use depends on the purpose of the map, the location of the subject area, the scale of the map, and the presumed audience. This selection process is a part of generalization, which is discussed later in this chapter. For example, a map showing climates of the United States would not normally call for transportation routes, unless the objective is to show a relationship between these features. In that case, the transportation net would become a part of the thematic information. The outline of the states and the location of major cities would be desirable for reference.

A frequent problem involves the amount of surrounding area that should be shown. Should only the outline of the area in question be drawn, or should the map extend beyond the borders of the subject area and include the neighboring areas (Figure 5.5)? Children sometimes assume a country ends where the map ends. Even adults, when confronted with an unfamiliar area, may need more background information to orient themselves. A very general rule of thumb is that, for a familiar area, such as one of the individual states of the United States (for American audiences), it is acceptable to show only the outline, but for an area assumed to be less familiar to the reader, such as Uruguay, Botswana, or the Czech Republic, some surrounding areas should be included as base information. Another way of handling this problem is to include an inset map that shows the surrounding area.

SOURCES OF INTERNAL BASE DATA

Often base data are taken from other maps, but locating appropriate base and reference maps can sometimes be difficult. If one is employed by a mapmaking agency, one can utilize the in-house map collection, but if one is an independent cartographer, one needs to utilize other sources. Obvious sources are sheet maps or atlases from government agencies, private companies, and institutions. One should also become familiar with local map sources such as university libraries and public libraries that have map collections. If such source maps are used in compilation, one must be aware of the copyright laws governing the use of published works. If a suitable base map is

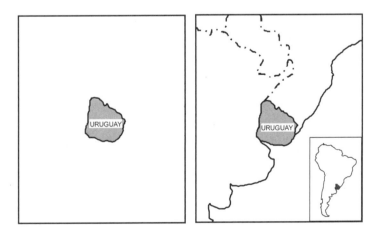

FIGURE 5.5. For an unfamiliar area some surrounding information is useful, as is an inset map.

not available, then one must use original surveys, remotely sensed imagery, or even construct a base from field surveys. Other sources are digital imagery and aerial photographs. Sheet maps and imagery can be scanned or digitized if they are not already available in digital form.

Much information is now available through the Internet in digital form. The USGS provides maps and outline files (sometimes called "shape files") online and many libraries have digitized map collections, especially of historic maps. Many websites sell digitized outlines. Software packages often include outline maps. When using any base map make sure to check the outline for accuracy against other maps. The author has found digitized outlines that place coastal cities 50 miles inland or that omit major islands and bodies of water. Such mistakes, the result of sloppy digitizing, are unfortunately common.

Primary Data

The primary or thematic data are the heart of the map. What is included here depends on the objective of the map. The term *thematic* is used in two different ways. In GIS, various layers are sometimes called "themes," so that one theme might be roads, another vegetation, and a third boundaries. For thematic maps, the *theme* is the subject of the map. Keep in mind that a thematic map should not normally illustrate more than one theme, although the theme may include more than one variable, such as education attainment related to income. It is better to make two or three maps than to overload a single map by trying to show too much information.

SOURCES OF PRIMARY DATA

It is not feasible to list the sources of all possible kinds of data here, but several categories may be examined:

- Field studies
- Imagery, including air photos, satellite imagery, radar, and the like
- Statistical sources
- Published maps
- Other printed materials
- Data on the Internet or on disk
- Interviews

Field studies include those published by various agencies, companies, and individuals as well as those that might be undertaken by the cartographer him- or herself. Frequently, maps created for journal articles, theses, dissertations, and the like are completely original, and their thematic information was compiled in the field by the author (who might also be the cartographer). Many books are available on fieldwork in geography.

Remotely sensed imagery ranges from black-and-white photographs taken from conventional aircraft to digital imagery relayed from satellites. This imagery is an important resource and should not be ignored. The subject is far too large to treat briefly or even as an entire chapter here. Anyone now planning a career in the mapping sciences should have a strong background in remote sensing, and even the casual cartographer, who makes maps only occasionally, is strongly advised to become familiar with the basics of the subject, either through course work or reading. (Several sources are listed in the Bibliography.)

Statistical data are commonly presented on thematic maps. The primary statistical sources for social data are the various kinds of censuses. The first regular national census was taken in Sweden in 1749. The United States produced its first census of population in 1790 and has taken a census every 10 years since. The early censuses recorded the names and ages of every member of the household, but this is no longer done. However, many more kinds of information are now obtained from the census of population and housing. In addition to the 10-year census of population and housing, censuses of agriculture and censuses of economics are also taken every 5 years. This information is online as well as in hard copy. The website *www.census.gov* is maintained by the U.S. Census Bureau. It is not always necessary to go to the full census publications. The *City and County Data Book* provides a summary of many kinds of information. If data are not needed on a block or tract basis, these summaries are helpful. For information on an individual state basis, the *Statistical Abstract of the United States* is invaluable. This annual publication provides reference sources for each table so that more detailed information can be found. The *Statistical Abstract* is also available online at the Census Bureau website. See Appendix B for a list of URLs for census information.

The Internet, of course, provides a wealth of statistical data. In addition to government censuses, there are also some private or commercial censuses or surveys. The latter are usually undertaken by special-interest groups, such as the Dairy Council or the California Wine Institute.

Other printed materials prove useful for some map topics. A helpful reference librarian can save hours of research time. Do not neglect books or articles on the map subject itself. Other maps may be available on the topic. Be aware that the laws of copyright apply: one must not copy or adapt, even by tracing, another map without permission and without giving credit. If no printed sources are available, interviews with experts, travelers, and others might be necessary.

Evaluating Sources

One must be concerned with the accuracy and reliability of sources. Since a map is normally compiled from a variety of sources, it becomes necessary to evaluate the credibility of each. If there is a difference in the accuracy of the sources used, the resulting map will be less reliable in some respects or in some areas than in others. One should, of course, use caution when using online sources; many websites created by individuals or even organizations have biases and don't provide complete information. *Know your sources.*

Copyright Ethics

It is unrealistic to expect all thematic maps to be completely original. It would be very expensive and time-consuming to construct a new base map for each thematic map or to construct the projection graphically or from tables. Obviously, thematic maps are built on the work of others. Therefore, some knowledge of copyright laws is necessary.

Publications and maps produced by the U.S. government are in the *public domain*, that is, they may be used without obtaining permission. This is not true, however, of publications produced by the private sector or by agencies in other countries. For example, British Ordnance Survey maps are produced under Crown copyright and a fee must be paid for their reproduction. Be aware that just because an organization has "national" or "American" in its name, it is not necessarily a government organization; the National Geographic Society and the American Automobile Association are examples. Permission must be obtained for using maps from either of these organizations even if they are only used for base data. If one is using source maps or thematic data that were taken from any copyrighted source, it is necessary to get permission in writing to use these data. A fee may be charged for permission; often, the wording of the source statement is specified. Do not assume that a map on the Internet is in the public domain. Copies of old maps published or printed by museums are also not necessarily in the public domain; the map itself may be beyond copyright date, but the copy or photograph created by the museum may be copyrighted.

Sources should be shown on the map in some way even for information that is in the public domain. This is akin to footnoting a statement in a written publication. The source statement serves the purpose of allowing the reader to go to the reference for more complete information and to judge the accuracy of the information. If your map is based on several other maps, but you show the information in a new way, you don't need permission, but you must acknowledge your sources on your map.

One should also keep a file of permissions and a record of sources, just as one compiles a bibliography for a written work.

GENERALIZATION

Because maps are drawn smaller than reality, they must be selective; not everything in the real world can be shown on the map, and the information that is shown cannot be portrayed exactly as it is on earth. Maps are *generalized*. *Generalization* refers to the selection, simplification, and even symbolization of detail according to the purpose and scale of the map. Both base data and thematic data must be generalized on thematic maps.

Generalization is required on all maps. While it is limiting to some degree, it should not be viewed as a negative factor. Generalization enhances the communicativeness of maps because it permits the subject of the map to stand out. It is not desirable to show everything on a map; to do so would merely result in visual clutter.

The generalization process is not simply a matter of leaving out some items and straightening rivers and coastlines. The number of articles and books about generalization in the literature makes this point obvious (see the Bibliography). Generalization is largely an intellectual process and requires the cartographer to develop a feel for generalization. The cartographer must have a thorough understanding of the purpose of the map and the nature of the item being generalized. Generalization is not a task that can be approached casually.

Goals of Generalization

The goals of generalization are to preserve geographic patterns and to stress thematic information. The cartographer strives for objectivity and uniformity of treatment in generalization. To accomplish these goals, the cartographer must be aware of the concepts of truth and accuracy as applied to maps. *Accuracy* is usually taken to mean locational or positional accuracy. Strict positional accuracy, however, is not always possible for thematic maps. Often, thematic maps are drawn at very small scales, so features cannot be drawn exactly to scale or located with precision.

For example, even at a comparatively large scale (for a thematic map) of 1:1,000,000, a major road 50-feet-wide would, if drawn to scale, be only 0.0006 inch or 0.0015 millimeter wide. Thus, to show both major and minor roads, the size must be exaggerated even though strictly speaking this is not accurate. Often it is desirable to show roads, rivers, and railroads. If these parallel one another through a narrow valley, which is a common occurrence, it becomes necessary to widen the valley. Otherwise, all three features cannot be shown. This also is not accurate. The USGS has established accuracy standards to solve such problems in generalization for topographic maps; this information is available at the USGS website listed in Appendix B. Similar standards can be devised for thematic maps.

It is possible to be truthful, however. If one accepts the idea that the map itself is a symbol and that the lines and other marks on it are symbols, not exact representations, then *truth* means showing the essence of patterns and relationships. Because the line representing a road in our first example is a symbol, it need not be a scale representation of the road. Because the relative position of the linear features in our second example is important, not the width of the valley, the fact that the valley is represented some feet wider that it really is presents no difficulty on a thematic map.

Operations of Generalization

Several techniques are used to create a truthful representation. Not all cartographers agree on all of these methods or operations, as Table 5.2 shows. Some have fewer methods because they combine operations, others break the operations into more categories. In this book I use eight operations: selection, simplification, smoothing, grouping, classification, exaggeration, displacement, and symbolization.

Selection

Selection is choosing what features will be shown on the map; it is involved in the generalization process in two ways:

1. *Choosing categories of data to be represented:* For example, one shows hydrographic features and roads, but not railroads. This aspect of selection is considered by some cartographers to be an activity separate from generalization.

2. *Choosing the amount of information within categories:* For instance, will only major rivers or only lakes of a certain size be shown, or will this map not be a true portrait of the hydrography?

TABLE 5.2. Operations of Generalization

Raisz	Robinson	Dent	Krygier & Wood	Slocum/Thibault	Buttenfield/McMaster	Tyner
Simplification	Selection	Selection	Selection		Selection	Selection
	Simplification	Simplification	Simplification	Simplification	Simplification	Simplification
	Classification	Classification				Classification
	Symbolization	Symbolization	Dimension conversion			Symbolizing
	Exaggeration		Enhancement	Exaggeration	Exaggeration	Exaggeration
			Displacement	Displacement	Displacement	Displacement
	Collapsing[a]			Collapse		
	Typification[a]					
Combination	Aggregation[a]			Aggregation	Combination	Grouping
Omission	Elimination[a]				Omission	
	Smoothing[a]	Smoothing		Smoothing		
				Merging		
				Amalgamation		
	Induction					
				Enhancement		
					Masking	
				Refinement	Emphasis	

[a]Robinson considers these subsets of selection, simplification, and exaggeration.

The selection process is not always simple. Selection is a matter of judgment and not easily computerized. Although some algorithms (procedures) have been developed for generalization, there are subtle decisions that cannot yet be specified precisely, quantified, or computerized. For example, if only perennial streams are represented, an arid area like Saudi Arabia might appear to have no drainage pattern.

Simplification

A complex feature, such as a coastline or river course, cannot be shown in complete detail on small-scale maps, nor is it desirable to do so. For most thematic maps (in contrast to topographic maps or nautical charts), the detail of the coastline is unimportant. There is no need to show each tiny inlet or point for most thematic maps, on a map of income, for example. The basic shape of the coast must be shown so that the area is recognizable, but not other detail. On some maps the nature of the coast—smooth, irregular, fjiorded—must be shown, but since it is impossible to be completely accurate and show every variation, many features are simplified even in these cases (Figure 5.6). The amount of simplification is a factor of scale; the larger the scale, the more detail that can be included.

Smoothing

Smoothing is a part of the simplification process. A road with many switchbacks is smoothed, as are rivers with many meanders (Figure 5.6). It is important that the character of the feature be preserved in the smoothing process, that is, the switchback road mustn't be drawn as a straight line and the meandering river must still show major changes in direction.

Grouping

Many small features are often grouped. In a simplistic example, individual trees are grouped into a forest. If the scale does not permit a number of small features to be

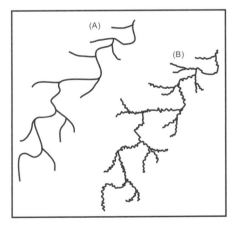

FIGURE 5.6. Features may be simplified and smoothed as in (A).

shown individually and they are close together, they may be grouped. Many maps created with GIS programs show great detail, but if the map is reduced for printing, the individual features coalesce into a "blob" and would be better grouped at the outset.

This is also the case when there are too many categories of a larger group to be shown, such as vegetation categories. Forty vegetation categories on a large-scale map might be useful to the researcher, but in an atlas they can be grouped into fewer broad categories (Figure 5.7).

Classification

We commonly classify or categorize data for many noncartographic purposes. A teacher will group a series of numerical exam scores into categories to assign grades; censuses group age categories and the like. Data for cartographic purposes are also categorized to make them more meaningful and to clarify or emphasize patterns of distribution. Just as a list of numerical scores in random order must be manipulated to make it meaningful, numbers spread over a map must be manipulated to make a significant display. Normally, this aspect of generalization applies to thematic data and is especially important when symbolizing volumes. The classification of data is discussed in more detail in Chapter 8.

Exaggeration

Earlier, the example of a road, a railroad, and a river in a narrow valley was put forth. If the scale of the map does not permit all three features to be shown without overlap, the valley can be widened or exaggerated to accommodate the features (Figure 5.8).

Displacement

When a road and a railroad parallel one another, they are frequently displaced, that is, moved farther apart so that the two lines can be distinguished (Figure 5.8). In the

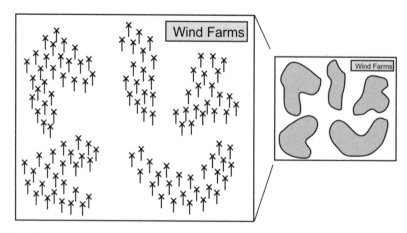

FIGURE 5.7. Data may be grouped into categories.

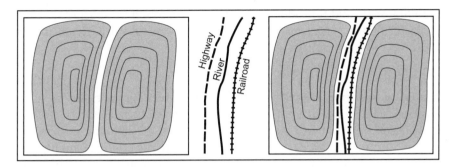

FIGURE 5.8. Exaggeration and displacement. The valley has been widened and the linear features compressed to show the three features in a narrow valley.

narrow valley example, the valley width may be exaggerated and the road, railroad, and river displaced.

Symbolization

Symbolization includes the selection and design of symbols to represent geographic phenomena on a map. Not all cartographers agree that symbolization is a part of generalization. Because assigning symbols to various phenomena is central to the cartographic process, some consider symbolization to be completely separate. In this book, symbolization, because of the size of the subject, is treated separately in Chapters 7 and 8.

Governing Factors

One approaches generalization differently for different maps. There can be no one rule for generalization that applies to all maps. Guidelines can be established, but they are governed by several factors: purpose of the map, scale of the map, the reader's skills and perceptual abilities, and the equipment and skills of the cartographer. Generalization is also affected by whether the information to be generalized is base or thematic data.

Data

GENERALIZATION OF INTERNAL BASE DATA

Internal base data are the information that forms a background for the thematic detail. The kinds of data are hydrography, boundaries, coastlines, and sometimes topography, transportation lines, and cities. These most often involve selection and simplification, but may also require exaggeration or displacement.

GENERALIZATION OF THEMATIC DATA

Thematic data most often have to be selected, categorized, and symbolized. Formulas have been devised to aid in choosing the number of items to be shown on a map based on compilation scale and reproduction scale, but, although this sort of quantification can be helpful, choices based on practice and experience are usually most valuable.

Topic and Purpose

The purpose of the map is probably the major governing or controlling factor of generalization. The same amount of generalization is not needed for rivers used as base information on a climate map as for rivers used as thematic information on a map showing drainage patterns, for example. Coastlines could be highly simplified on a map of population, while a small-scale, general-purpose map would need more detail, and a map designed for navigation must show the coastline in great detail. The map's purpose is especially important in choosing what will and will not be shown on the map. The cartographer must give considerable thought to the topic and purpose of the map before generalizing.

Scale of the Map

All things being equal, the smaller the map scale is, the more generalized the map must be. For the same purpose, topic, and audience, a large-scale map needs less simplification and can show more detail than a small-scale map of the same subject. This is the major limiting factor. One must keep in mind both drawing and reproduction scales. If the linear dimension of the map is to be reduced by one-half, generalization at the drafting scale will be greater than if the map will be reproduced at the same size (Figure 5.9).

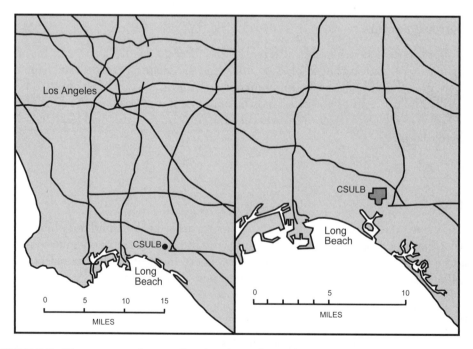

FIGURE 5.9. The amount of generalization depends on the topic, purpose, and scale.

Reader's Abilities

In all cartographic activities the reader must be considered. Although the cartographer does not always know the specific skills of the audience, some perceptual abilities are considered universal. For example, the normal human eye cannot distinguish more than six shades of gray between black and white on a map. Therefore, it is pointless to use 12 different shades of gray to illustrate 12 different categories on a statistical map. Either a different method of symbolization must be found or the data must be grouped to form fewer categories.

The eye cannot perceive very fine lines or read lettering smaller than about 0.04 inch; therefore, to use such fine lines or small letters defeats the communication purpose of the map.

Environmental factors must also be considered. As discussed in Chapter 2, the requirements for a map that is to be viewed in the low-light conditions of an airplane cockpit are different from those for maps that will be viewed in a brightly lighted room; a map designed for an automobile GPS screen has different requirements than those of a paper road map. These factors affect the degree of simplification, the number of categories, or the type of symbolism. Sometimes it is necessary to perform tests with focus groups to determine the amount and kind of generalization needed.

Cartographer's Equipment and Skills

Although most professional maps and many amateur maps are produced with the assistance of computers, the kind of equipment used and the skills of the mapmaker are still important. Screen, printer, and scanner resolution all affect the amount of detail that can be shown or printed. Software capabilities are an important factor. Programs vary widely in their design and generalization usefulness. Much software is updated on a nearly annual schedule, so the mapmaker needs to be aware of its current status.

While drafting skills aren't important for working with GIS, illustration, or CADD programs, skill at using the programs is. It takes time to master any software program. While one might produce a simple map without having complete mastery of a program, one requires more advanced skills to create more sophisticated maps.

Quality of Data

The quality and nature of the data must also be taken into account. Imprecise or poor data do not permit one to make a precise map. The data may not be proper for the kinds of symbols the cartographer wants to use. For example, a quantitative map cannot be made from qualitative data. If the data form is inappropriate, it is necessary either to find suitable data or to change the map type. One must also be aware of bias. Unless he or she is making a map designed to persuade, the cartographer should strive to be objective. One must realize that data from special-interest organizations might be highly subjective. Bias can also enter the map through selection of the phenomena to be mapped, the data used from a set, or the control points used.

SUGGESTIONS FOR FURTHER READING

Bielstein, Susan M. (2006). *Permissions, a Survival Guide: Blunt Talk about Art as Intellectual Property*. Chicago: University of Chicago Press.

Buttenfield, Barbara, and McMaster, Robert B. (1991). *Map Generalization: Making Rules for Knowledge Representation*. Harlow, UK: Longman Scientific and Technical.

Fitzsimons, Dennis E. (1985). Base Data on Thematic Maps. *American Cartographer, 12,* 57–61.

McMaster, Robert B., and Shea, K. S. (1992). *Generalization in Digital Cartography.* Resource Publications in Geography. Washington, DC: Association of American Geographers.

Monmonier, Mark. (1996). *How to Lie with Maps* (2nd ed.). Chicago: University of Chicago Press.

Steward, H. J. (1974). "Cartographic Generalisation: Some Concepts and Explanation." *Cartographica*, Monograph 10.

Chapter 6

The Earth's Graticule and Projections

> ". . . but then I wonder what Latitude or Longitude I've got to?"
> (Alice had not the slightest idea what Latitude was, or Longitude
> either, but she thought they were nice grand words to say.)
>
> —Lewis Carroll, *Alice in Wonderland* (1865)

THE EARTH'S GRATICULE

The Size and Shape of the Earth

Many aspects of mapmaking would be greatly simplified if the earth were a plane surface. However, the earth is not flat, as was realized early in the history of cartography. Attempts were made even by the early Greeks to represent the actual surface with as little distortion as possible.

The true figure of the earth is not a regular shape like a sphere or an ellipsoid, but a lumpy solid called a *geoid*, which simply means "earth-shaped figure." As Table 6.1 shows, the polar radius is roughly 13 miles less than the equatorial radius. This does not seem like a great difference (on a 12-inch globe the difference is only 0.019 inches or 0.05 centimeter) but for large-scale maps, such as topographic maps or navigation charts, the difference is significant. In addition to flattening of the poles there are other smaller irregularities.

Geodesists, who deal mathematically with the shape of the earth, must consider actual earth measurements, as must surveyors and cartographers who make topographic maps and GIS specialists who make large-scale city maps and the like. For most thematic mapping, however, we are not concerned with the irregularities of the earth's shape, since we commonly make maps of very small areas or small-scale maps for which geodetic accuracy is not a major issue.

TABLE 6.1. Earth Dimensions

Feature	Kilometers	Statute miles
Equatorial diameter	12,752.401	7,926.677
Polar diameter	12,713.45	7,899.988
Equatorial circumference	40,074.788	24,902
Meridional circumference	40,007.198	24,860

Latitude and Longitude

It is somewhat more difficult to locate points on a spherical surface than on a plane surface since a sphere has no convenient starting and ending points. It is necessary to establish a reference system for the earth in order to locate places with precision. Fortunately, because of its motion, the earth does have two fixed points, the north and south poles, which can be used to establish a reference system. The earth spins on an imaginary axis, the ends of which are the poles, and the north pole always points to the same spot in space.

A *great circle* (one that divides the earth into two equal hemispheres) drawn midway between the poles and at right angles to them is called the *equator* and serves as the starting line for the reference system. Distances from the equator are measured in degrees north or south to the poles. The angular distance north or south of the equator is called *latitude* and is measured from 0° at the equator to 90° north and 90° south at the respective poles (Figure 6.1).

If we connect all points having a given latitude, a *small circle* is formed parallel to the equator at the distance of the chosen latitude. This small circle is called a *parallel*. A parallel is a line on the globe that joins all points having the same angular

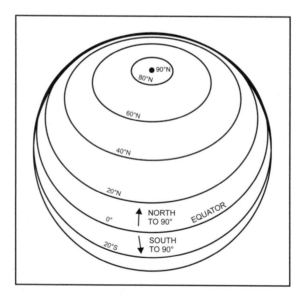

FIGURE 6.1. Parallels are lines of equal latitude.

distance (latitude) from the equator. *Parallel and latitude are not synonyms; a parallel is a specific line, and latitude is a distance.*

An infinite number of parallels can be drawn, since degrees can be divided into minutes, seconds, and fractions of seconds. This allows us to locate any place on earth precisely with respect to the equator. If the earth were truly spherical, the distance in miles between each degree of latitude would be the same. However, as we have seen, the earth is not a perfect sphere, and thus there is some variation in the length of a degree of latitude. For most small-scale mapping, this variation is too small to be significant.

The system of parallels is only one-half of the earth's reference system. There are 360° along each parallel, and to pinpoint a location, we must also specify where it lies on the parallel. Unfortunately, there is no fixed point or line on the earth comparable to the poles or the equator that can be used as a convenient origin for measurement along the parallels. For many years each country used a true north–south line passing through its capital or some other significant location. Distances were measured from this line in degrees and called *longitude*. When maps of only one country were used there were few problems; but with faster travel around the earth and the use of a variety of maps, this system became cumbersome. Many atlases in the United States during the 19th century contained maps on which the longitude from Washington was printed along one border and the longitude from London along the opposite border to eliminate tedious comparisons and calculations.

In 1884, the International Meridian Conference established the line through the transit instrument at the Royal Observatory at Greenwich, England, as the starting line for east–west measurement; this line is called the *prime meridian*. The angular distance east or west of the prime meridian to some other point, measured from 0° to 180°, is the longitude. If a series of lines is drawn at right angles to the parallels and connecting the poles (by passing planes through the poles), the lines are true north–south lines and are arcs of great circles. These lines are called *meridians*. A meridian is a line that joins all points having the same longitude (Figure 6.2). Note that each meridian is one-half of a great circle. Also note that like parallel and latitude, meridian and longitude are not synonymous. Meridians are lines, longitude is angular distance. The grid formed by parallels and meridians is the *graticule*.

Distance

Although the terms *distance* and *direction* are used frequently in everyday speech, there is confusion about the actual meaning of these terms when referring to the earth's spherical surface.

On a plane, the shortest distance between two points is a straight line, but on the surface of a sphere, the shortest distance between any two points is an arc of a great circle. Thus, if two points are located on the same parallel, the shortest distance between them is not along the parallel (unless that parallel is the equator), but along the great circle joining them. This circle can be found on a globe by stretching a string between the two points.

A convenient unit for measuring distance on a globe is the *nautical mile*. A nautical mile is defined as 1 minute of arc on a great circle of a sphere that has the same surface area as the earth. Given the scale of most globes, this can be taken as simply

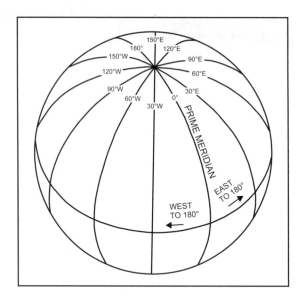

FIGURE 6.2. Meridians are lines of equal longitude.

1 minute of arc on a great circle. We determine the number of minutes separating the two points by comparing the distance to the same distance on the equator of a globe. If it is necessary to have this distance in statute miles or kilometers, it can easily be converted (see Table 5.2).

Direction

Direction is a difficult concept when applied to a sphere. *Direction* is the position of one point on the earth relative to another point. It is usually measured by the angle between a reference line and the shortest line that can be drawn from the point of observation and the point of interest. Since the shortest distance on a sphere is a great circle, then the direction of a point on the earth is the angle between the reference line (a meridian) and an arc of the great circle running between the observer and the point. Direction is stated as *azimuth*. Azimuth is the angle measured clockwise from the reference line and has a value between 0° and 360° (Figure 6.3).

Compass directions such as north, south, east, west, southwest, and so on are commonly used when speaking of direction. A line of constant compass direction, that is, a line that cuts all meridians at the same line, is called a *loxodrome* or *rhumb line*. An aircraft following a constant compass course other than north or south would trace a spiral toward a pole, but would theoretically never reach it. This is a *loxodromic curve* (Figure 6.4).

In this book "loxodrome" or "line of constant compass direction" refers to these directions, and the term "azimuth" is used synonymously with "direction." It is important to distinguish between these terms because, as we shall see later, some projections show lines of constant compass direction as straight lines, and others show great circles as straight lines. These are mutually exclusive except for the meridians and the equator.

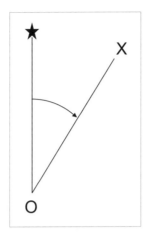

FIGURE 6.3. Azimuth is the angle between a true north–south line and a great circle passing through the observer and the point observed.

Other Grids and Coordinates

For some purposes, grids other than latitude and longitude are used. The primary ones in the United States are the Universal Transverse Mercator Coordinate System, the State Plane Coordinate System, and the Public Land Survey System. USGS topographic maps show all three of these systems in addition to latitude and longitude.

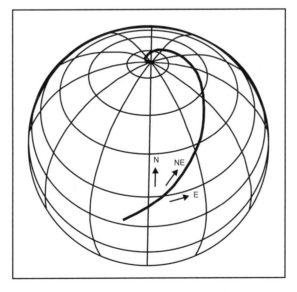

FIGURE 6.4. A loxodromic curve (rhumb line) crosses each meridian at the same angle. It spirals toward the pole.

Universal Transverse Mercator Coordinate System

The Universal Transverse Mercator Coordinate System (UTM), like latitude and longitude, is used to pinpoint locations on the earth. The UTM grid was developed by the U.S. Army in the 1940s and is based on the transverse Mercator projection. The system divides the earth into 60 zones that are 6° of longitude wide. Each zone is centered on a meridian. The zones are numbered from 1 to 60 beginning at 180° and going eastward. Thus, Zone 1 goes from 180° to 174°W and is centered on 177°W (Figure 6.5).

Each longitude zone is divided into 20 latitude zones. Each of these zones is 8° from north to south. Latitude zones begin at 80°S and extend to 84°N. The last zone is extended 4° so that the northern-most lands are covered. The latitude zones are given letters beginning with "C" in the south and continuing to "X" in the north. The letters "I" and "O" are not used because they can be confused with the numbers one and zero. "A," "B," "Y," and "Z" are used for portions of the Antarctic and Arctic areas. The grids formed are referred to by the longitude number and the latitude letter with longitude first. Thus, Los Angeles is in Zone 11S and New York is in Zone 18T.

A place on earth is pinpointed by the longitude zone and an *easting* and *northing* coordinate pair. Easting is the distance of the point from the central meridian of the longitude zone and northing is the distance of the point from the equator. Distances are in meters.

State Plane Coordinate System

The State Plane Coordinate System (SPS or SPCS) shown in Figure 6.6 is widely used by state and local governments and is considered more accurate than the UTM sys-

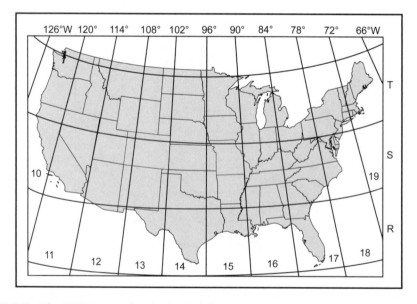

FIGURE 6.5. The UTM zones for the United States.

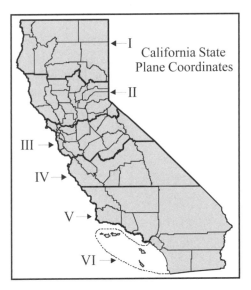

FIGURE 6.6. State Plane Coordinate System for California.

tem because the zones are smaller. Because the zones are small and because they are based on states the SPCS is not useful for national and regional mapping.

The SPCS, like the UTM, is based on zones, but these generally follow political boundaries so the map of zones appears more irregular. The number of zones in a state depends upon the size of the state; the boundaries within a state follow county lines except in Alaska. The SPCS uses about 120 zones to cover the United States. Like the UTM, eastings and northings are used, which are distances east and west of the origin. The origin is usually 2,000,000 feet west of the zone's central meridian.

U.S. Public Land Survey System

The U.S. Public Land Survey System (PLSS) was first proposed by Thomas Jefferson and implemented shortly after the end of the Revolutionary War. Two ordinances, the Land Ordinance of 1785 and the Northwest Ordinance of 1787, provided for systematic rectangular survey of federal lands so that those lands could be transferred to private citizens. The PLSS is a *cadastral* or property-mapping system. Not all states are covered in the PLSS because it provided for survey before settlement. Thus, the original colonies are not covered, and Texas has its own system (Figure 6.7). The PLSS was initially begun in Ohio, but because the system was not yet fully developed, the surveys in Ohio and Indiana are different from those of other covered states. There are also some areas where other cadastral systems were in use, such as Louisiana, which used the French *arpent* system, or long lots; New England, which used the New England town system; and areas of the Southwest, which used Spanish systems.

The PLSS is based on 37 named principal meridians and base lines, as shown in Figure 6.7 that provide origins. True north–south lines (meridians) and east–west lines are drawn from these origins at 6-mile intervals called "tiers and ranges" (while

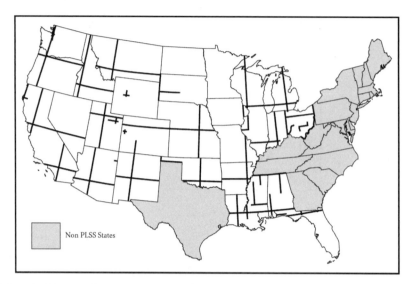

FIGURE 6.7. Public Land Survey System. The original 13 colonies and Texas are not covered by the PLSS.

"tiers" is the original term, common usage has become "townships and ranges"). The 36-square-mile areas formed are called *townships* (Figure 6.8). Each township is divided into 36 one-square-mile areas called *sections* that contain 640 acres, and the sections can be further subdivided into quarter sections, eighth sections, and so forth so that any parcel of land has a unique one-line designation. For example, NE1/4,SW1/4,SW1/4,NE1/4, sec. 23, T1SR3W, Mt. Diablo Meridian would describe a 2 1/2-acre piece of land located in the first tier south of the base line, the third range west of the Mt. Diablo principal meridian in the northeast quarter of the southwest quarter of the southwest quarter of the northeast quarter of section 23.

The PLSS is a systematic cadastral survey, but in much of the world unsystematic property surveys, which are also called *metes and bounds* or *indiscriminate location,* have been used. For metes and bounds, property lines and boundary markers are arbitrarily selected and are often natural features such as rocks, rivers, or trees. A property description might begin, "From the granite boulder at the river bank N30°E, 500 yards to the oak tree, then S80°E 150 yards. . . ." These boundaries, unlike the PLSS, were usually drawn after settlement and the lots tend to form a crazy-quilt pattern.

TRANSFORMING THE EARTH'S GRATICULE: MAP PROJECTIONS

A projection is the transformation of the earth's spherical graticule onto a plane surface. The formal definition is *a systematic and orderly representation of the earth's grid upon a plane.* Unlike some shapes, such as a cone, a sphere cannot be flattened without some distortion. The projection is a foundation because it governs the positions of points on the map with respect to one another. The projection forms the framework for the map, and therefore it is necessary to decide which projection will be used early in the planning stages of the map. There are an infinite number of projections possible, and more than 400 projections available, although only a relative few are employed regularly.

The key terms in the definition are "systematic" and "orderly." One does not create a "projection" simply by peeling the globe or tearing off pieces. The graticule is transformed or projected in a systematic manner onto a flat surface, and the positions of continents and seas then plotted onto the transformed graticule.

The shape of the earth was discussed in the preceding section. Here we will not be involved with the geoid or spheroid, but will consider the earth to be a sphere for ease in understanding. Although for large-scale maps, such as topographic maps, the exact shape of the earth must be taken into account, for most thematic maps the earth can be considered a perfect sphere.

Before analyzing projections, the properties of the globe should be reviewed:

1. All parallels are parallel.

2. Meridians are half great circles and converge at the poles.

3. The meridians are evenly spaced along any parallel.

4. Quadrilaterals formed by the same two parallels and having the same longitudinal dimensions will have the same areas.

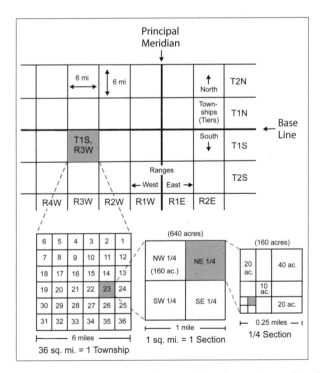

FIGURE 6.8. Townships and sections. Any place within the PLSS can be identified by township and section.

5. Parallels are equally spaced along the meridians (assuming the earth to be a sphere).

6. Area scale is uniform; that is, 1 square inch of the globe in the high latitudes and 1 square inch in the low latitudes covers the same earth area.

7. Distance scale is uniform.

Only a globe can combine all these properties because its shape is virtually the same as the earth's. When the spherical graticule is transformed onto a plane, which radically changes its geometry, unavoidable distortions are introduced. Therefore the only true representation of the earth's surface geometry is a globe (Figure 6.9).

It must not be assumed, however, that because flat maps have distortions they are merely poor substitutes for globes or that globes are without disadvantages. Globes are limited by their practical size. A globe 14 inches in diameter has a scale of 1 inch to 566 miles, hardly feasible for navigation or showing roads. Globes permit only one-half of the earth to be seen at a time. Many thematic maps illustrate global patterns; in such cases, it is desirable to view the entire earth at once. A climate map, for example, is more useful if the entire world pattern can be seen at once and regions compared. Finally, to show all of the thematic maps in even a small atlas would require many globes, which would present storage and handling problems.

Projections can be problem-solving tools and should not be considered negative devices. The distortions of flat maps can be used to advantage. Phenomena that cannot be shown on a globe can be presented on a flat map by exploiting projection properties. For example, the gnomonic projection shows all great circle arcs as straight lines so it can be used for plotting the shortest distance between two points. The Mercator projection shows loxodromes as straight lines and can be used to plot

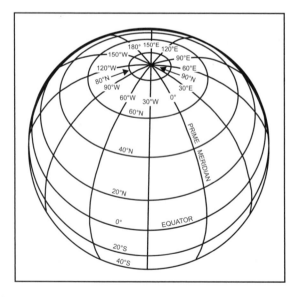

FIGURE 6.9. Earth's grid.

FIGURE 6.10. The cordiform or Stab–Werner projection. "The Cartographer's Valentine."

constant compass routes useful for navigators. A projection can also permit the cartographer to zero in on an area and present it in large scale.

There is no single best projection, although sometimes claims are made for new projections that state they are better than any previously constructed projection. This simply is not true. All projections have distortions, and a new projection merely has distortions that are different from those of previously used projections. One cannot even say there is a best projection for a particular use. From the infinite projections possible, there are usually several that could be used for any mapping project, although if enough conditions are specified, the choice can be limited.

While there are an unlimited number of projections and variations on projections possible, many of those already devised are not in common use but remain cartographic curiosities (Figure 6.10). Only about 25 or 30 projections and their variations are widely used (see Appendix A).

The discussion that follows is a simplified and nonmathematical description of projection concepts. Several excellent sources are listed in the Bibliography for those who want a more rigorous treatment.

Classification of Projections

Many attempts have been made to classify projections. Unfortunately, it is not possible to devise mutually exclusive categories since there is always some overlap. The most common classifications are based on *projection properties* and *projection surfaces*.

Projection Properties

It is desirable to preserve some of the geometric properties of the earth's grid when creating or using map projections. These basic properties and the projection types that preserve them are as follows:

1. *Equivalence of area.* Maps that preserve area scale are called *equal-area* or *equivalent* projections.

2. *Preservation of angles or of shape of small areas.* Projections that have this proporty are called *conformal* or *orthomorphic* projections.

3. *Linear scale.* Projections that preserve the linear scale for some part of the map are called *equidistant* projections.

4. *Direction, specifically azimuths.* These projections are called *azimuthals* or *zenithals.*

There are various other properties, such as showing great circles as straight lines and the like, but these have no specific category. A last category is a *compromise* group that has no special properties, but presents a good general appearance that is not terribly distorted in any of the properties.

EQUIVALENT PROJECTIONS

A uniform area scale is maintained on equivalent or equal-area projections. Equivalence is obtained by matching stretching with compression. That is, if there is stretching in one dimension, there must be equal compression in the orthogonal (at right angles) dimension. To do this, angles may be distorted, which means that shape is sacrificed. The shapes in Figure 6.11 all have the same area. It is possible to maintain equivalence throughout the projection.

CONFORMAL PROJECTIONS

These are also called *orthomorphic* projections. Often conformal projections are incorrectly described as true-shape projections, but the shape quality is limited. Conformal projections preserve angles with infinitely short sides; therefore, only small features, such as a bay or peninsula, retain correct shape; larger areas, such as countries or continents, are distorted in overall configuration.

Conformality is achieved by meeting two conditions: (1) parallels and meridians must cross at right angles, and (2) scale must be the same in all directions about a

FIGURE 6.11. These shapes each have the same area.

FIGURE 6.12. While the figures in each column have the same shapes, their areas are different.

point. That is, if there is stretching in the longitudinal dimension, it must be matched with an equal amount of stretching in the latitudinal dimension (Figure 6.12). As a result, angles are preserved at the expense of a severely distorted *areal* scale, and squares of 100 square miles in different parts of the globe, while remaining square, will have different apparent sizes on a conformal projection. Figure 6.13 shows a conformal world map compared to an equal-area map. Note the size of Greenland on these two maps. Compare the shape and size to that found on a globe.

Because of the conflicting requirements of equal-area and conformal projections, it is not possible to preserve both properties on a flat map; *it is impossible to design an equal-area conformal projection*. These properties can be achieved simultaneously only on a globe.

EQUIDISTANT PROJECTIONS

A map is called equidistant if distances are shown correctly. Equidistance is not attainable over the entire map; the distance scale of a map is correct only along certain lines or from specific points.

AZIMUTHAL PROJECTIONS

Projections that show azimuths correctly are called *azimuthal* or *zenithal* projections. Unlike conformality and equivalence, this property cannot exist everywhere on the map. *Azimuth is correct only from a single point, the center*; it is not possible to measure the azimuth between any other points on the map. A map that shows constant compass direction is *not* an azimuthal map.

COMPROMISE PROJECTIONS

These are also called *minimum error projections*, a somewhat misleading term. These projections have no particular property except appearance. These are not conformal, but shapes may not be terribly distorted; they are not equal area, but areas may not be

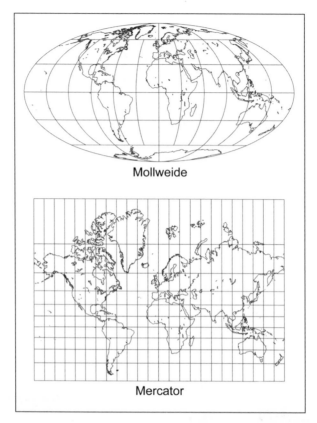

Mollweide

Mercator

FIGURE 6.13. The Mollweide projection is equal area and the Mercator is conformal. Compare the size of Greenland and South America on the two projections.

poorly represented (Figure 6.14). This is a very useful group of projections for much thematic map work.

Projection Surface

A second common way to classify projections is by the projection surface. Surfaces that can be cut and flattened without distortion or tearing are called *developable surfaces*. The sphere is not a developable surface; it cannot be flattened without distortion or tearing. If, however, the graticule is transformed onto a surface that is developable, that surface can be flattened without any additional distortion (Figure 6.15). A cone can be placed on a globe, the graticule projected onto it, and the cone then unrolled into a plane. The cylinder and the plane may also be used as projection surfaces; the cylinder can be flattened and the plane is already flat. The cylinder and plane may be thought of as extreme cases of a cone. A cone that touches the equator and has therefore an infinitely high apex is a cylinder, and as cones are placed on higher and higher parallels, they approach the plane.

Projection, in theory, is a two-stage process. First, an imaginary globe is created at the intended scale. The imaginary globe is called the *generating globe* or *reference globe.* The scale of the map projection is called the *defined scale* or the *nominal scale* and is the same as that of the generating globe. The second stage is to project the globe onto a plane.

If we imagine one of the developable forms placed on a transparent reference globe with a light source inside, we can then visualize the earth's graticule being projected onto the developable surface. The name of each group of projections is taken from the surface upon which the map is projected. Consequently, we have cylindrical, conic, and plane projections. A final group of projections is made up of those that cannot in any way be imagined as projected onto one of these surfaces. Various names are given to these last, the most common being *mathematically devised* or *conventional projections.*

Of course, projections are not normally created by a light source and a globe; this would be done only for demonstration or teaching purposes. They are created mathematically and drawn by a computer or, in the past, manually constructed geometrically or from tables (Figure 6.16). Only a limited number of projections can actually be projected by light or by geometric projection (these are called *perspective projections*); but if a projection has the general appearance characteristics of a cylindrical projection, for example, it is placed in this class.

Deformation

As we have seen, all projections have deformations. The larger the area represented, the more significant those distortions become. For example, for a map showing a very small area, such as that of a 1:24,000 topographic map or a city street map, angular deformation, scale differences, and area distortion are negligible. However, if one tries to join topographic maps to cover a larger area, one quickly realizes that there are gaps between the sheets; joining all margins produces a curved surface. If the

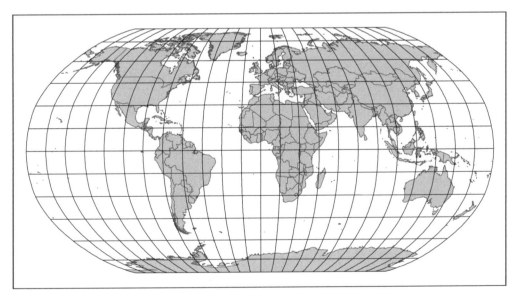

FIGURE 6.14. The Robinson projection, created in 1963, is a compromise projection.

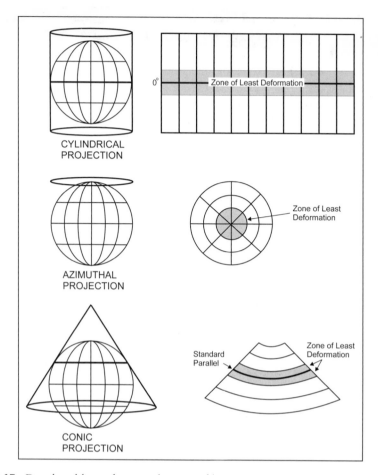

FIGURE 6.15. Developable surfaces and zones of best representation.

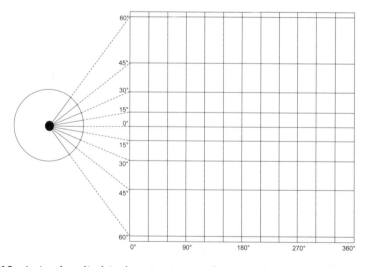

FIGURE 6.16. A simple cylindrical projection can be created geometrically.

entire earth is represented on a flat map, the distortions become obvious and trouble-some. Since many thematic maps are of countries or the entire earth, the selection of a projection becomes of paramount importance. The cartographer must choose the projection with the most advantages for the problem at hand and one for which the distortions are least intrusive.

There are several ways of analyzing projection distortions. Tissot's indicatrix, which permits quantification of distortion, is frequently used. It is also possible to draw *isoanamorphic lines* (lines of equal distortion) on a projection to show distor-tions graphically. Because this chapter is a nonmathematical introduction to projec-tions, I will take a more intuitive, if less rigorous, approach.

Scale varies from point to point on any projection, and at any point the scale may vary in different directions. The variations can be measured by a *scale factor* (SF). The SF is equal to the scale on the reference globe at the point divided by the defined scale. If SF = 1.0, there is no distortion.

Distortion is not random on projections; it follows a definite, orderly pattern for each projection. On every projection there is a point (or points) or line(s) that is (are) correct, that is, the SF is 1.0. These are called *standard points* and *standard lines*. For a short distance away from the point or line, there is an area of minimum distor-tion. This is called the *area of least deformation* or the *zone of best representation*. By matching this zone with the area to be represented and considering the desirable properties, the cartographer can choose a suitable projection. In the discussion of projection types that follows, the zone of best representation for the various groups will be designated. Figure 6.15 shows the pattern of these zones for the major catego-ries of projections.

Cylindrical Projections

If a cylinder of the same diameter as the generating globe is wrapped around the equa-tor of the globe, and the graticule is projected onto the cylinder, a simple cylindrical projection results (Figure 6.15). When the cylinder is unrolled, the shape of the paper is a rectangle, the meridians appear as straight, parallel, equally spaced lines all the same length, and the parallels appear as straight, parallel lines all the same length as the equator. Maps that have this basic appearance are called *cylindrical projections* and vary only in the spacing of the parallels. The meridians on a globe or the earth converge at the poles, but on cylindrical projections, since the meridians are parallel to one another, there is east–west stretching that increases toward the poles. They are also called *rectangular projections*.

Parallels on the earth are small circles and are smaller in circumference than the equator (a great circle), but on a cylindrical projection all parallels are the same length. This again shows the east–west stretching of these projections. Since the cyl-inder is wrapped around the equator in the *normal case*, the circumference of the cylinder is the same as the length of the equator. Therefore, the scale is correct along the equator and increases poleward. It can be seen that the zone of best representation is along the equator.

Since scale is true only along the equator on these projections, it is customary to label the map "equatorial scale 1:xxx,xxx,xxx" or to use a graphic scale devised especially for cylindrical projections. Since each parallel has a specific ratio with the equator, a graphic scale for each parallel can be constructed. Figure 6.17 illustrates such a scale for a cylindrical (Mercator) projection with an equatorial scale of 1:1,000,000.

There are many cylindrical projections. Although they all have a family resemblance, they may be constructed to preserve different properties. For example, there are cylindrical equal-area, cylindrical conformal (Mercator), and several compromise cylindrical projections. These projections vary only in the spacing of the parallels.

A subgroup of the cylindricals that is sometimes described are the *pseudocylindrical* projections. These projections have straight parallels with the meridians equally spaced, but the meridians are curved.

Probably the best known projection is the Mercator projection (Figure 6.18), which is a cylindrical conformal projection. It is also one of the most misused of the projections. Gerardus Mercator (Gerard Kramer) created the projection in 1569 to solve a particular mapping problem. The 16th century was a time of exploration and yet a period when few navigational aids were available to sailors. The compass existed and the tools to determine latitude, but longitude was difficult to determine at sea, and great circle routes were hard to follow. The navigator needed a chart that would allow him to make a landfall, despite his relatively poor instruments. Although a short route was desirable, finding port was more important. Mercator's projection was designed to show constant compass directions (rhumb lines) as straight lines, which means that a simple protractor could be used to read and plot compass angles (ship's course) correctly (Figure 6.19). Because it was conformal, the shapes of small features, such as bays and harbors, were correctly represented. This was a major breakthrough for navigation at the period.

Unfortunately, the Mercator projection has been much abused. Because of its rectangular shape and the neat appearance of the graticule, because it is easy to draw and fits nicely on a page, it has been widely used for maps of the world, especially for wall maps in schools. Many significant misconceptions were produced in the minds of generations of students because of this projection's severe distortion of area. Commonly, students would wonder why Greenland was an island, but South America was a continent. Because areal distortion is especially great in the high latitudes, Greenland appears to be about the same size as South America, whereas it is actually about the size of Mexico. In addition, since many people assume that a straight line between two points is the shortest distance on any map, erroneous ideas of distance result. Figure 6.20 shows the relationship between a rhumb line and a great circle

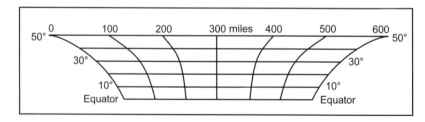

FIGURE 6.17. Scale for a cylindrical projection.

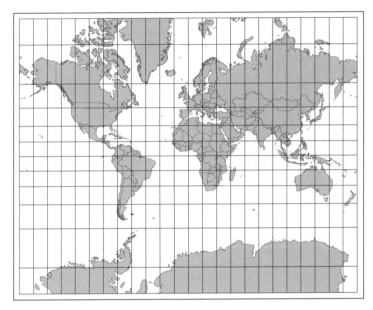

FIGURE 6.18. The Mercator projection was developed by Gerardus Mercator in 1569 for navigation.

route between Los Angeles and London. On this projection, the great circle route appears longer than the rhumb line.

The Mercator–Peters Controversy

Cartographers have objected to the use of the Mercator (or any cylindrical) projection for world maps for many decades, as described earlier. However, in the late

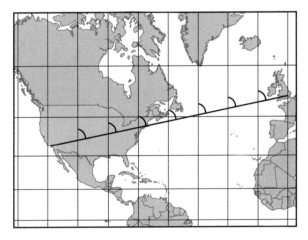

FIGURE 6.19. Rhumb lines or loxodromes are straight lines on the Mercator projection. The cardinal directions, N, S, E, W, are thus shown as straight lines.

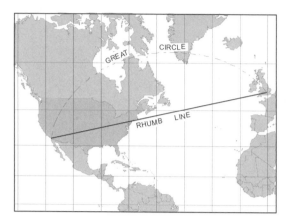

FIGURE 6.20. On the Mercator projection a great circle that is the shortest distance between two points on a sphere looks farther than the rhumb line. These distances can be compared on a globe.

1970s, Arno Peters proposed a "new" world map, the so-called Peters projection (Figure 6.21), saying that it was a far superior projection than Mercator, which he believed was used because cartographers were biased toward developed countries. His projection, which is actually the Gall projection developed a century earlier, is an equal-area cylindrical projection, and, according to Peters, shows shapes better than Mercator. In fact, his projection has many deficiencies of its own, some of which are inherent in any cylindrical projection. A number of agencies adopted this projection based on Peters's persuasive rhetoric and it is now found in atlases and wall maps.

Because of the distortions inherent in rectangular projections, in 1989 the

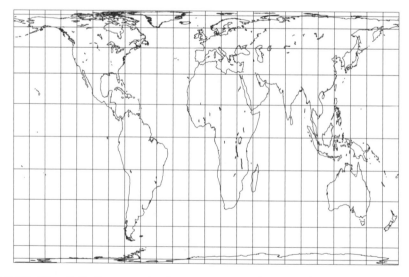

FIGURE 6.21. The Gall–Peters projection was proposed as a replacement for the Mercator, but has the same disadvantages of any cylindrical projection.

American Cartographic Association passed a resolution strongly urging publishers, governmental agencies, and the media to cease using rectangular world maps for general purposes and artistic displays, which was signed by other cartographic and geographic organizations (American Cartographic Committee on Map Projections, 1989, p. 223). Despite this effort, rectangular maps are still found in children's geography workbooks.

Other Aspects of Cylindrical Projections

If it is possible to wrap a cylinder around the equator, it is equally possible to create a cylindrical projection by wrapping the cylinder around some other great circle. If the great circle is formed by a pair of opposing meridians, the resulting projection is a *transverse aspect*. For any other great circle, the projection is an *oblique aspect*. The parallels and meridians look very different in these aspects from the normal or *equatorial aspect*, and some properties are lost. For example, rhumb lines are straight lines only on the equatorial aspect of the Mercator, not on the transverse or oblique aspects. The zone of best representation is along the tangent great circle in any of these cases (Figure 6.22).

It is also possible to create a cylindrical projection that has two zones of good representation, called a *secant case*. This is accomplished with a secant cylinder, that

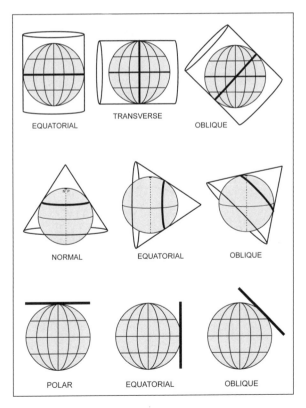

FIGURE 6.22. Projection aspects.

is, a cylinder that cuts the globe instead of encircling it. The small circles where the cylinder enters and exits the generating globe has true representations.

Plane (Azimuthal) Projections

If a plane is placed tangent to the pole of the generating globe and the earth's graticule projected upon it, a *plane projection* is the result (Figure 6.15). Such a projection will be circular in outline and have straight meridians that radiate from the center and parallels that are concentric circles. The spacing of the parallels and the amount of the earth that can be shown will be governed by the position of the imaginary light source. The tangent point or center can be located at the poles, at the equator, or any point between the equator or pole. These create *polar, equatorial,* and *oblique* aspects of the projections. Obviously, in the equatorial and oblique aspects, parallels will not look the same as on the polar version, but the various projection properties will be maintained. Some of the plane projections, such as the equidistant, are used more often in oblique aspects than polar.

Since all plane projections have the property of showing azimuths correctly from the center point, these are more commonly called *azimuthal projections.* Any straight line that passes through the center of these projections is a great circle. Another term sometimes encountered for this group is *zenithal projections.* Since the plane touches the globe at the center, the zone of best representation for azimuthals is at the center point. Keep in mind that the zone of best representation is the point where the plane is tangent, and is only at the pole in the polar case.

There are five azimuthal projections in common use (Figure 6.23): the ortho-

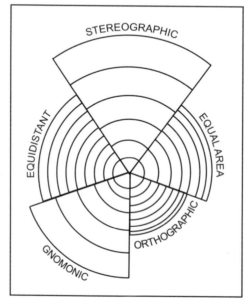

FIGURE 6.23. The five azimuthals in polar case vary only in the spacing of the parallels. Redrawn from Robinson, Arthur H., et al. (1995). *Elements of Cartography* (6th ed.). New York: Wiley. Adapted with permission from John Wiley & Sons, Inc.

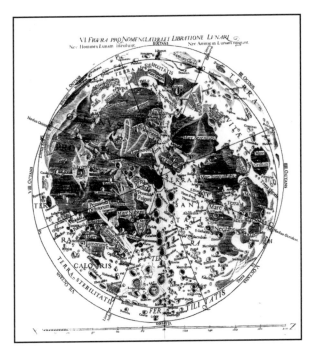

FIGURE 6.24. The moon was often shown in orthographic projection on early lunar maps (Riccioli, 1651).

graphic, the stereographic, the gnomonic, the azimuthal equal-area, and the azimuthal equidistant projections. The appearance of these projections varies only in the spacing of parallels (in the polar case). All can be constructed by geometric projection.

Orthographic Projection

If a light source is assumed to be at infinity, the light rays appear to be parallel. An azimuthal projection produced in this way is called an orthographic projection. We know that Ptolemy used this projection, which he called the "analemma," for representing the heavens, and may have developed it. The orthographic was also a frequently used projection for early maps of the moon because when the moon is viewed from the earth, it appears to be on the orthographic projection (Figure 6.24).

Figure 6.25 shows that an orthographic projection cannot present more than one hemisphere at a time. When centered on the pole, it has parallels that become closer together as the equator is approached. Other than azimuths being correct from the center and great circles passing through the center being straight lines, both of which are true for all azimuthal projections, the orthographic has no outstanding properties save appearance. The orthographic closely resembles the view we have when looking at a globe, and therefore it is often used when one simply wants to create a good visual appearance.

FIGURE 6.25. The orthographic projection in oblique case.

Stereographic Projection

If the light source is assumed to be at the opposite point on the globe from the tangent plane, the stereographic projection is created. Here it is possible to show a little more than one hemisphere, but less than the entire sphere. Normally, however, only a hemisphere is shown; to represent the entire earth a pair of hemispheres is used.

In Figure 6.26, it can be seen that the parallels and meridians cross at right angles and that the spacing of the parallels in pole-centered cases become greater as one approaches the equator. As one goes outward from the center of the projection, east–west stretching is introduced. On the stereographic projection, the spacing of the parallels increases in the same proportion as the spreading of the meridians. Thus, this is a conformal azimuthal projection.

On the stereographic projection, circles on the globe appear as either circles or arcs of circles unless a great circle arc passes through the center of the projection—these appear as straight lines.

Gnomonic Projection

If a light source is imagined to be in the center of the globe, the gnomonic (pronounced *no mon' ic*) projection is created. This appears to be the oldest projection used, since it was known to Thales of Meletus about 600 BCE and he is usually given credit for its development. On this projection, it is impossible to show an entire hemisphere, and the distortion of shapes and areas is extreme (Figure 6.27). Despite these disadvantages, it is a widely used projection because of one important quality: all great circles appear as straight lines, and all straight lines on the map are great circle arcs on the gnomonic projection. It is, therefore, of importance to navigators because great circle routes can be easily plotted.

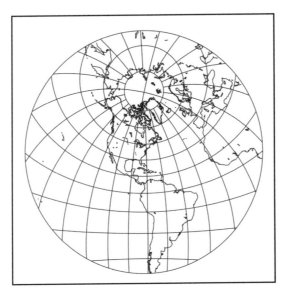

FIGURE 6.26. The stereographic projection in oblique case. The stereographic is a conformal azimuthal projection.

In navigation, the gnomonic projection is used with the Mercator projection to plot courses. Since the compass heading changes constantly along a great circle course, too many corrections are required to fly or sail a true great circle course. On a Mercator the compass heading does not change along a straight line, but this line normally deviates significantly from a great circle route. It is possible, however,

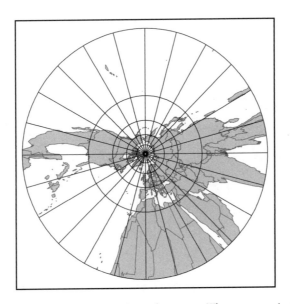

FIGURE 6.27. The gnomonic projection in polar case. The gnomonic has distortion that increases away from the center, but all great circles are represented as straight lines.

to chart a great circle route by plotting it onto a Mercator Projection and following short rhumb-line segments as shown in Figure 6.28. The gnomonic projection is used to find the true great circle and the Mercator to plot the short legs.

Not all azimuthal projections can be visualized as created by projecting a transparent globe by means of a light source. Two important azimuthals are mathematical projections: the azimuthal equal area and the azimuthal equidistant.

Azimuthal Equal-Area Projection

Since east–west stretching is produced by the failure of meridians to converge poleward on the plane projections, compressing must be introduced to get equivalence. On *Lambert's azimuthal equal-area projection* (Figure 6.29), this is done by placing the parallels closer together the farther away they are from the center. Almost the entire earth can be shown on this projection. Although the azimuthal equal-area projection cannot be projected with a light source and transparent globe, it is possible to construct it graphically.

On this projection, azimuths are shown correctly in common with others of the azimuthal family, but areas can also be compared.

Azimuthal Equidistant Projection

Another commonly used projection is the azimuthal equidistant, which has parallels truly spaced along the meridians (Figure 6.30). This projection is most commonly centered on a place other than the poles to enable one to determine distances from the center to any other place in the world and to determine the starting azimuths from those places. Distances between any other places will not be correct; distance can only be measured from the center point.

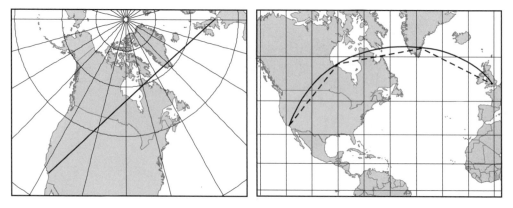

FIGURE 6.28. Great circle routes plotted on the Gnomonic can be approximated with short rhumb lines on the Mercator.

FIGURE 6.29. Lambert's azimuthal equal-area projection in oblique case. The azimuthal equal-area shows sizes of places correctly although shapes are increasingly distorted away from the center.

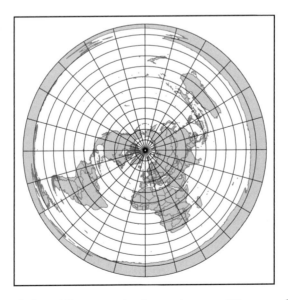

FIGURE 6.30. Azimuthal equidistant projection, polar case. Distances from the center and only the center can be measured accurately. The projection can be centered on any point on the earth. It is frequently used to show airline distances.

Conic Projections

The conic projections, in theory, are created by placing a cone on the generating globe so that it touches a parallel. This parallel is called the *standard parallel of projection* and its length is the same as the parallel on the generating globe. When the cone is flattened, it forms a fan shape, with straight, radiating meridians and curved parallels that are concentric arcs. It is not possible to show the entire earth on such projections (see Figure 6.15).

As with other projections, the zone of best representation is in the area where globe and surface touch, in this case, along the standard parallel, which is thus analogous to the tangent point on azimuthal projections, or to the tangent great circle on cylindrical projections. Distortion increases away from this standard parallel, making conic projections most suitable for midlatitude areas of greater east–west than north–south extent.

It is possible to create conic projections that have two standard parallels by making a cone that cuts the globe. In such secant cases, the projection has two zones of best representation, which are along the lines where the cone intersects the globe. By proper selection of these parallels, the two zones may be brought close enough together to form a single wide zone of little deformation. Two of the most commonly used projections are secant case conics: Lambert's conic conformal and Albers's conic equal area.

Lambert's Conic Conformal

For Lambert's conic conformal (Figure 6.31), the parallels frequently chosen are 33° and 45°. This makes the projection especially suitable for representing the United States. Although this is a conformal projection and, by definition, not equivalent, areal distortion over the United States is small (5% maximum), and the linear scale error is only 2.5% at a maximum. This projection is also commonly used for aeronautical charts.

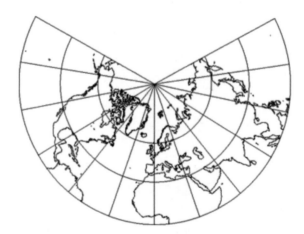

FIGURE 6.31. Lambert's conic conformal. The entire earth cannot be shown on conics and distortion increases away from the standard parallels.

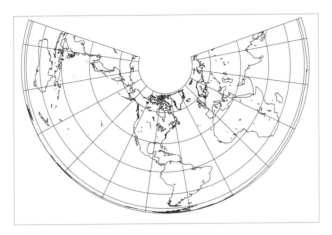

FIGURE 6.32. Albers's conic equal-area projection.

Albers's Conic Equal Area

The conic equal-area projection (Figure 6.32) is also much used for the United States, having standard parallels at 29° 30' and 45° 30'. Despite being an equivalent projection, shapes are represented well, and for the United States the maximum linear error is only 1.25%.

Other Forms of the Conic

It is possible to place the cone on a small circle that is not a parallel. If a cone is placed so that its axis does not coincide with that of the globe, oblique and transverse aspect conic projections are formed.

An interesting variation of conic projections is the *polyconic*. It might be reasoned that, if a single cone produces one zone of best representation, and a secant cone two zones, it might be possible to create more zones by a series of standard parallels. If one visualizes a series of cones of differing sizes, each touching a different parallel, then one could "peel off" these different standard parallels and place them together along a central meridian (Figure 6.33). To eliminate the gaps between the different parallels, it is necessary to introduce some stretching, but along the central meridian where the strips touch there is a north–south zone of good representation. When extended over the entire earth, this projection is very distorted, as can be seen in Figure 6.32, but for narrow areas of great north–south extent it is very accurate. The polyconic was the projection used for USGS topographic maps until recently.

Mathematical Projections

Mathematical projections include the many that cannot even be imagined as produced by a transparent wire globe and a light source. The shapes of these projections vary widely: oval, circular, heart-shaped, star-shaped, and even armadillo-shaped. Some appear to have been "peeled" from the surface of the globe, although exami-

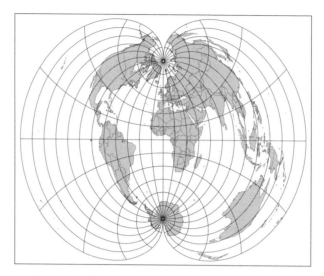

FIGURE 6.33. The polyconic is formed of a series of standard parallels. It doesn't show the entire earth well and is usually used only for small areas, such as topographic maps.

nation reveals that this is not possible. Large numbers of mathematical projections have been developed, especially in recent years, and many are designed for a specific purpose. Many, although by no means all, are equal area. Only four of these will be discussed here for purposes of illustration.

Sinusoidal Projection

The sinusoidal projection, which is also called the Sanson–Flamsteed projection, appears to have been first used by Nicholas Sanson about 1650. This is an equal-area projection that has a straight central meridian and straight, equally spaced, true-to-scale parallels. The central meridian is also true to scale. This makes it possible to measure distance along any parallel (not great circle distances) and along the central meridian. A zone of least distortion is produced along the equator and central meridian (Figure 6.34). Any meridian may be chosen as the central meridian. The projection gets its name from the meridians being trigonometric sine curves. Shapes toward the outer margins of the projection in the high latitudes are badly distorted, but for areas in the center of the projection, such as Africa or South America, the sinusoidal projection would be a good choice.

Mollweide Projection

In some ways, the Mollweide or homolographic projection (Figure 6.35) resembles the sinusoidal. Both are equal area and show the entire earth, and both have a straight central meridian and straight parallels. The Mollweide, however, is an ellipse lacking the pointed poles of the sinusoidal. Only the 40th parallels north and south are correct in length, and the parallels are not spaced truly on the central meridian. Shapes are not as badly distorted in the polar areas as they are on the sinusoidal. On this

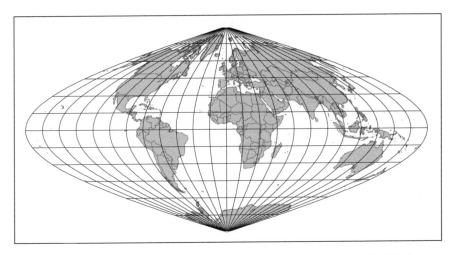

FIGURE 6.34. The sinusoidal projection. This is an equal-area, pseudocylindrical projection.

projection, the zones of best representation are around the central meridan and the 40th parallels.

Goode's Homolosine Projection

In 1923, J. Paul Goode devised the homolosine projection by combining the sinusoidal and homolographic projections. The homolosine is made up of the sinusoidal from 40°N to 40°S and the homolographic from the 40th parallels to the poles, thus combining the "best" parts of the two projections and extending the zone of

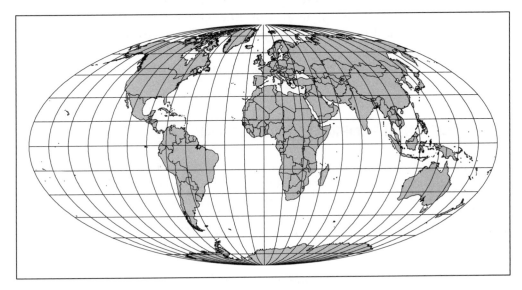

FIGURE 6.35. The Mollweide projection (homolographic).

good representation (Figure 6.36). The projection remains equal area. It is rarely seen except in the interrupted form (see below).

Robinson Projection

Arthur H. Robinson in 1963 devised his projection in an effort to create a world projection that had a pleasing appearance with few obvious distortions (see Figure 6.14). This projection is a *pseudocylindrical* and a compromise in that it is neither equal area nor conformal. It has become quite popular.

Interrupted Projections

With projections of the type of the Mollweide and the sinusoidal, it is possible to interrupt the graticule and thereby create a series of zones of best representation. To illustrate this idea, the Goode's homolosine is used as an example (Figure 6.36). Although this example uses manual construction for explanatory purposes, in practice the projection would be created on a computer.

A line is constructed to represent the equator and marked in the desired number of degrees for the meridians. A central meridian is chosen for each continent and drawn to its proper length. Points are chosen for each interruption. Now the sinusoidal graticule is constructed around each central meridian to the marked interruptions. Each continent has been placed within the zone of best representation for the projection. Note that nothing has been removed at the points of interruption, that certain meridians have been repeated, and that the entire projection is made up of a series of portions of the sinusoidal. This process can be carried out with any projection that has this essentially oval outline. Interruptions may be placed on either land or ocean depending on the area of interest, and any meridian may be chosen as the central meridian.[1]

Interrupted projections can also be *condensed*. Condensed projections are interrupted projections that have had an area of little interest removed to allow a larger scale map to be placed on a page (Figure 6.37). When the focus is on land areas, a portion of the Atlantic Ocean is frequently removed. When projections are condensed, there should be some indication on the map that this has been done; otherwise a misleading map is created.

CHOOSING AN APPROPRIATE PROJECTION

Although an unlimited number of arrangements of the earth's graticule is possible, only comparatively few are in common use. The cartographer should be aware of the advantages and limits of at least the most widely used of these.

Although map projections are now created by computer rather than by manual means, this does not relieve the mapmaker of the task of choosing an appropriate

---

[1]Another way of creating an interrupted projection is by joining a series of azimuthal projection segments, as in the star-shaped projection.

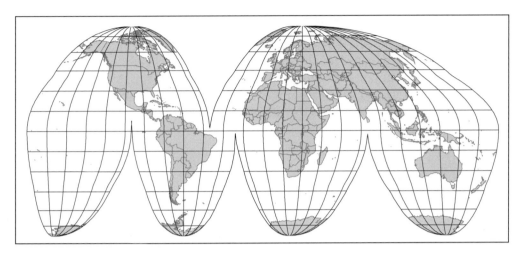

FIGURE 6.36. Goode's homolosine projection is a combination of the sinusoidal and Moll-weide (homolographic) projections. It is almost always in interrupted form.

projection from those available. Even if the computer program provides a list of suitable projections for a given task, the cartographer must be able to make an intelligent choice among them. While two or more projections may be suitable, probably not all will illustrate or communicate the given situation equally well. A well-chosen projection can enhance the communicative value of the map; a poorly chosen projection may even mislead the map reader.

To make the best choice, it is necessary to consider the purpose of the map, the subject of the map, the size and shape of the subject area, the location of the subject area, the audience, and the size and shape of the page. Other considerations are the appearance of the graticule, the general attractiveness of the projection, and the availability of the projection within the chosen software program.

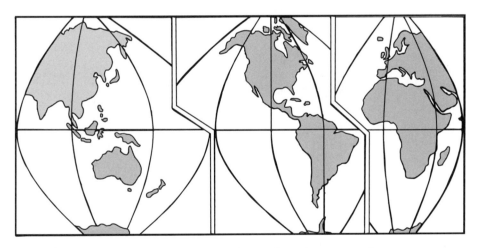

FIGURE 6.37. An interrupted projection can be condensed to gain scale.

Subject and Purpose of the Map

The subject and purpose of the map are the most critical factors for picking a projection. For distribution maps, equal-area projections should be used; for maps showing world distributions, a nonequivalent projection will usually be misleading since quantity and area are related (Figure 6.38). When the distribution is illustrated by dots, the impression of the relative density of the distribution will be distorted if nonequivalent projections are used.

For navigation, maps that show angles or azimuths properly are needed; still other needs might dictate equidistance or conformality. Appendix A will aid in the selection of a suitable projection. The USGS has created a poster of map projections and their properties (see Appendix B).

Size and Shape of the Area

As we have seen in the previous discussion, some projections are best suited for areas of great east–west extent, others for north–south areas. Reversing these would result in serious distortion within the area mapped. The area of interest should always be placed in the zone of least deformation. Therefore the cartographer should be familiar with those zones for the commonly used projection groups.

For small areas (large-scale maps), the choice of projection is seldom of any consequence, since at this scale the differences between projections are negligible. A map of a city or county will look essentially the same on any projection. (There are occasions when projection must be taken into account for small areas, but these are usually of more concern to surveyors than to thematic mapmakers.)

For world projections, the size and shape of the area will determine if an inter-

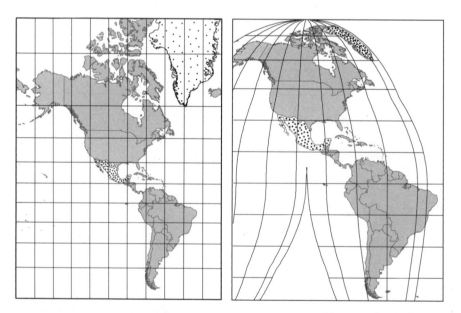

FIGURE 6.38. Greenland and Mexico have the same number of dots, but the conformal projection gives an erroneous impression of density.

rupted projection should be used and where the interruptions should be placed. Obviously, if the distribution being mapped extends over both land and water, an interrupted projection will not be suitable; if the distributions are confined to land, placing interruptions in the ocean areas would be appropriate.

Location of the Subject Area

Just as the zone of best representation of the various projections makes them more suitable for areas of different sizes and shapes, so too are they suited to different latitudinal positions. For example, the conic projections generally show midlatitudes best and especially midlatitude areas that have a greater east–west than north–south extent. An oblique conic or a portion of the polyconic projection could be used for narrow north–south areas.

If one is using a world projection, the issue of centering arises. Normally, one puts the primary area of interest in the center of the map, but for a world distribution map on which no one area is primary, one must decide what to put in the center. Projections in normal or equatorial aspects can be centered on any meridian. Commonly the prime meridian is chosen. However, some have accused cartographers of being deliberately Eurocentric when using the prime meridian because it places Europe in the center of the map. However, studies have found that when students create mental maps of the world, they tend to put their home country or continent in the middle so centering for the audience makes sense. Thus, a reasonable option is to place North America in the center for western hemisphere audiences or Asia for Chinese audiences and so forth. Again, some will criticize the selection, but if the cartographer is aware of potential bias, has considered the audience and purpose of the map, and based the choice on these factors, the choice can be defended.

Audience

Map audiences vary widely, but, as detailed in Chapter 2, maps are generally aimed at a specific audience. When the cartographer chooses a projection, he or she has some knowledge of the degree of sophistication of the map user. Taking this information, the cartographer can decide if the graticule form is distracting or if an interrupted projection might be confusing.

Appearance of the Graticule

The graticule as shown on different projections varies enormously in appearance. In most cases, one does not want a graticule that is so complex and eye-catching that it dominates the page. A very complex graticule might also be confusing for an untrained reader.

Size and Shape of the Page

Since cartographers are often restricted to given page dimensions, this aspect of the format can be a definite factor in choosing a projection. Sometimes a clever choice of projection will allow one to place a map within awkward dimensions. It is also possible to condense an interrupted projection to gain scale and put a larger scale map on a specific page size.

Orientation Indication

Frequently one sees maps in textbooks, atlases, and other works that do not show the projection name. This is not a problem when the area represented is a small one, such as a city or county, but for larger areas, omitting the projection name does the reader a grave disservice. This is especially so when the graticule has also been omitted, leaving only ticks to show latitude and longitude, or when even the ticks have been left off (a practice that cannot be recommended for world maps). When this is done, the reader has no way of determining where the zone of best representation is found or what the projection properties are.

Except for maps of easily recognizable areas where no reason exists for determining latitude and longitude and where the orientation is obvious, some indication of orientation is needed. The graticule will serve this purpose, as will grid ticks.

North arrows are the most frequently misused indicators of direction, as noted in Chapter 2. A north arrow does not show the general direction of north on maps; it is specific. The north arrow must be true for the entire map, not just one small spot; therefore, unless the meridians are all straight and parallel, a north arrow cannot be used. This means that, except for maps of small areas, the use of such an arrow is limited to cylindrical projections in the equatorial case. It is not suited to conics, azimuthals, or any projection with curved meridians.

Misuse of Projections

As we saw for the Mercator projection, map projections are often misused. The Mercator projection is an elegant and valuable projection, perfectly suited to its purpose when used for navigation; it has been given a bad name because it was used inappropriately. On maps designed for propaganda purposes, the misuse is deliberate and intended to mislead the reader. In other cases, projections are misused through ignorance, inexperience, or negligence.

Projections can be misused or abused in several ways. Incorrect choice of projection is a common error—for example, choosing a non-equal-area projection to illustrate a distribution or other area-dependent topic, such as size comparisons or countries. Using a projection that shows true distance only along the equator is a poor choice for a world geography textbook that includes map questions asking students to determine distances between midlatitude cities.

Another practice that creates misleading maps is the omission of any indication of graticule. Without a graticule (or tick marks) for reference, the reader cannot judge the properties of the projection. Since graticule omission is frequently combined with omission of the projection name, the reader has no way of knowing what can and cannot be shown correctly on the map. In some especially disturbing instances, a completely different frame has been placed around the projection. Richard Dahlberg illustrated an interrupted projection that had the lobes "filled in" by extending Antarctica across the gaps in the projection (Figure 6.39).

A commonly found abuse is the use of the so-called geographical projection (Fig-

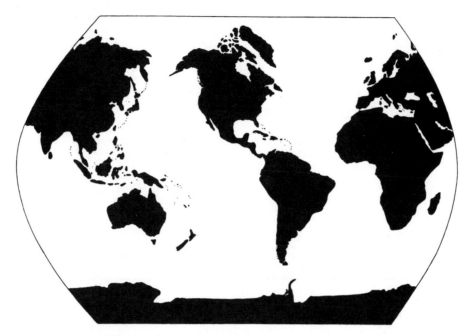

FIGURE 6.39. Map projections are often abused. In this case an interrupted projection has had the grid removed and Antarctica spread across the gaps between the lobes. From Dahlberg, Richard E. (1961). Maps without Projections. *Journal of Geography*, *60*, 213–218. Reprinted with permission from Taylor & Francis.

ure 6.40). It is created by using longitude and latitude as rectangular or *x*, *y* coordinates and is similar to plane charts of the 16th century. Mark Monmonier flatly states, "If any projection is worth denouncing, it's the vaguely named 'geographical projection' popular among users of geographic information system (GIS) software" (Monmonier, 2004, p. 176). On the map shown, one can see that the map enlarges the northern states of the United States; this is because, like other rectangular projections, the meridians do not converge at the poles.

Other examples of the misuse of projections abound and yet should not happen. There is little excuse for ignorance given the number of readily available tables of projection use, and no excuse for negligence or laziness. Using an unsuitable projection because it was the most readily available base map is not acceptable.

SUGGESTIONS FOR FURTHER READING

Bugayevskiy, Lev M., and Snyder, John P. (1995). *Map Projections: A Reference Manual*. London: Taylor & Francis.

Greenhood, David. (1964). *Mapping*. Chicago: University of Chicago Press.

Johnson, Hildegard Binder. (1976). *Order Upon the Land: The U.S. Rectangular Land Survey and the Upper Mississippi Country*. New York: Oxford University Press.

Kennedy, Melita, and Kopp, Steve. (2004). *Understanding Map Projections*. Redlands, CA: ESRI Press.

FIGURE 6.40. The so-called geographical projection is plotted on a rectangular latitude–longitude grid. It has no particular properties and is highly distorted.

Monmonier, Mark. (1995). *Drawing the Line: Tales of Maps and Cartocontroversy.* New York: Holt.

Monmonier, Mark. (2004). *Rhumb Lines and Map Wars: A Social History of the Mercator Projection.* Chicago: University of Chicago Press.

Price, Edward T. (1995). *Dividing the Land: Early American Beginnings of Our Private Property Mosaic.* Chicago: University of Chicago Press.

Snyder, John P. (1993). *Flattening the Earth: Two Thousand Years of Map Projections.* Chicago: University of Chicago Press.

Snyder, John P., and Voxland, Philip M. (1989). *An Album of Map Projections.* U.S. Geological Survey Professional Paper 1453. Washington, DC: U.S. Government Printing Office.

Thrower, Norman J. W. (1966). *Original Survey and Land Subdivision: A Comparative Study of the Form and Effect of Contrasting Cadastral Surveys.* Washington, DC: Rand McNally and The Association of American Geographers.

PART III
SYMBOLIZATION

Chapter 7

Basics of Symbolization

> Cartography is about representation.
>
> —ALAN M. MACEACHREN, *How Maps Work* (1995)

Maps are a form of graphic communication that like other graphic forms convey meaning primarily by symbols. Symbols are the graphic language of maps; the selection and design of symbols are a major part of creating a successful map.

To represent spatial concepts, relationships, and distributions, such as the pattern of population density in the United States, in words is extremely cumbersome and generally ineffective; the reader simply cannot grasp the relationship between area and distribution. A verbal description must be read in linear fashion; that is, a sentence must be read from start to finish to make sense, not from the middle outward. But a map can be read in any order and still be understood. In fact, with a simple map, the meaning can be assimilated almost instantly.

Since many thematic data are statistical, it might seem that a tabular array of statistics would display the data just as effectively as a map, but as Table 7.1 and Figure 7.1 show, a table, like a verbal description, cannot convey a map's spatial component. Even if numerical data are displayed on a map in their correct geographic arrangement, the reader will not receive the immediate impression of relationships that a map with appropriate symbols provides.

Cartography is not the only discipline that uses symbols. Mathematicians and musicians use symbolic notation that can be understood by others in their field anywhere in the world. Some have suggested that maps, because of their symbolic nature, are a form of universal language. This suggestion has only limited validity, however, because there is no standardized system of symbols for thematic maps. Such systems have been proposed, but it appears that the adoption of any one system is unlikely to occur in the near future. There is a degree of standardization for topographic maps—contour lines are used for elevations, and many other symbols are common to maps of various countries. Thus, while the maps may have a somewhat different appearance

TABLE 7.1. Vermont Cattle and Calves, 2007

County	Number
Addison	62,263
Bennington	3,370
Caledonia	13,550
Chittendon	10,469
Essex	5,550
Franklin	62,636
Grand Isle	5,857
Lamoille	6,365
Orange	18,239
Orleans	37,918
Rutland	15,673
Washington	7,161
Windham	6,526
Windsor	9,246

overall, a user can read a Swiss, a German, an English, or an American topographic map with little difficulty. There are standard symbol sets for certain other series of maps; for example, there has been some recent work on standardizing symbols for emergency response maps that could be understood by readers of any background.

Maps are a complex form of symbolic communication. Some study and effort are needed by the map reader to become fluent in the graphic language. Understanding international traffic signs, which are essentially pictographs, requires only a single level of perception. One need only recognize the symbol to know its meaning—Stop!,

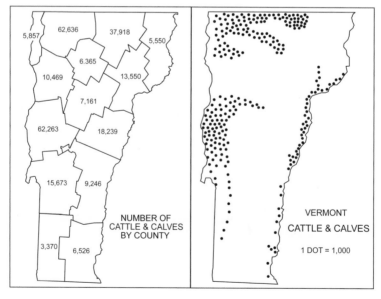

FIGURE 7.1. Tables and even numbers placed on a map are not as effective at showing a spatial distribution as symbols.

Basics of Symbolization   133

Yield!, Do not enter!—no further analysis is necessary, and no legend is required on the sign. This is basically symbol matching, not interpreting (Figure 7.2). It is this kind of symbolization that is proposed for emergency response maps. To interpret a map, however, the reader must not only identify the symbols from the legend, but also recognize the spatial relationships between the objects symbolized; this is a three-level problem in perception: identify symbols, recognize patterns, and then interpret patterns.

Because of this multilevel perception, the cartographer must choose symbols that are distinct and easily identified in the legend and must also choose the symbol system that most effectively portrays the relationships featured. Well-chosen symbols result in a map that is readily understood because it states its message clearly and unambiguously. A poor choice of symbolism produces a map that is confusing and may even mislead the reader.

We must distinguish between two terms that are sometimes used synonymously: *phenomena* and *data*. *Phenomena* are the features and attributes of the real world that are being mapped. They are the features to be conceived and communicated, such as rivers, roads, and countries, and the attributes of those features, such as form, continuity, and location.

Data are facts gathered by measuring, counting, calculating, or derivation. Geographic, or spatial, data are those that describe or measure aspects or attributes of geographic phenomena. It is important to remember that we are portraying spatial data, rather than phenomena, on maps.

In this chapter the nature of the geographic phenomena and data being symbolized, the measurement of data, and the nature of symbols are discussed so that an appropriate symbol can be chosen.

FIGURE 7.2. Qualitative pictorial symbols. These symbols require little or no interpretation and may be grasped quickly.

THE NATURE OF GEOGRAPHIC PHENOMENA

Anything that has a spatial component can be mapped, whether it is a tangible, visible feature on earth, a concept, or even an opinion. If a phenomenon varies with respect to location, it constitutes geographic reality and can be displayed in map form.

There are three spatial attributes of geographic phenomena: form, continuity, and location.

Form

Geographic or spatial phenomena can be thought of as existing at points, along lines, over areas (or polygons as they are usually described in the GIS literature), or as volumes (Figure 7.3).

Point phenomena exist at discrete points and may be actual or conceptual. Telephone poles, houses, and fire hydrants are all point features. The location of a city can also be viewed as a point when reduced on a small-scale map even though it has area in the real world.

Linear phenomena are features that are line-like in reality and may be thought of as having only one significant dimension: length. Some are tangible, such as rivers or roads. Others, such as political boundaries, cannot normally be seen on the earth's surface. Still others are conceptual and derived from information over time, such as average traffic volume on a highway.

Areal phenomena have two dimensions, length and width, and are distributed over a defined area. Like point and linear phenomena, they may be either directly observable or conceptual. Examples include vegetation types, land-use patterns, and living preferences.

Volume phenomena, like areal phenomena, extend over areas, but they are conceived of as having a third dimension. This dimension is a value or quantity, such as elevation of the land, inches of rainfall, or millibars of barometric pressure. Other examples are population density and percent of land devoted to crops. While elevation is directly observable as a three-dimensional phenomenon, the other examples represent a conceptual three-dimensional surface called the *statistical surface*. When data are collected there is an assumed x, y (or latitude, longitude) location and if the data are quantitative, there is also a z dimension. The z-value is the quantity represented and stands for height on the statistical surface.

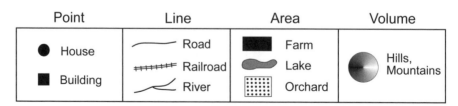

FIGURE 7.3. The forms of spatial phenomena.

Continuity

Geographic phenomena may be discontinuous or *discrete*, that is, found only at certain locations, or *continuous*, that is, found everywhere within the mapped area. Point phenomena are by nature discontinuous, but area and volume phenomena may be either. Barometric pressure and temperature are continuous over the world; forests are not, although they may be considered continuous over small areas. Discrete phenomena may also be either concentrated or dispersed.

Geographic Location

The geometry of geographic space is determined by reference to coordinates, either longitude and latitude for conventional mapping, or some other form for cartograms or other distorted maps.

LEVELS OF CLASSIFICATION OF GEOGRAPHIC DATA

The data gathered for geographic phenomena may simply show location and nature, that is, where and what the phenomena are; or they may rank phenomena, or they may specify actual values. There are four levels of measurement for data, each of which provides a certain degree of information. These are *nominal, ordinal, interval,* and *ratio* classification. Nominal classification is also called *qualitative* and is the lowest level of information. Ordinal, interval, and ratio classification come under the general heading of *quantitative* information and represent successively higher levels of information (Figure 7.4).

Nominal classification locates and names. Items are placed into mutually exclusive categories and located on the map. The categories are qualitative, since no ranking or value is attached, and no item can be placed in more than one category. Thus, a vegetation map shows deciduous forests, coniferous forests, grasslands, and the like, and a land-use map might identify urban and rural categories. Where there is some mixture, such as a forest with 90% deciduous trees and 10% coniferous trees, the category would be chosen as a part of the generalization process (see Chapter 5).

Ordinal classification "orders" or places data into ranked categories, but does not give exact values within categories. Sometimes a series of numerical categories is given, while at other times a descriptive ranking, such as small, medium, or large is used. A city might be classified as 500,000–1,000,000 population, but no indication of its exact place within the category is given.

Interval and ratio classification are similar, since actual values are given, as well as function and rank. A uniform unit of measurement is used for both, such as degrees, feet, meters, Euros, or dollars. They differ in that interval measurement has an arbitrary starting point instead of an absolute zero point. The Celsius temperature scale begins at an arbitrarily assigned zero, the temperature at which water freezes. Population measure begins at an absolute zero. Therefore, comparisons such as twice as hot or half as hot are not possible using interval measurement, but twice as many people or half as many people is possible with ratio measurement. In cartographic practice, the difference between these two levels of measurement has little or no effect

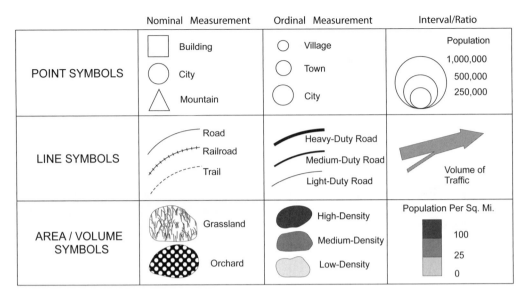

FIGURE 7.4. Classification of geographic data.

on symbol selection. When symbols are attached to data that have been scaled by interval or ratio measurement, value can be determined from the size or some other characteristic of the symbol.

VISUAL VARIABLES OF CARTOGRAPHIC SYMBOLS

Symbols on a map[1] show the position, the nature, and sometimes the value of phenomena. All symbols have location on the map, which corresponds to the position of the object. Because maps are two-dimensional, only three kinds of marks may be used to symbolize data. The three marks are points, lines, and polygons; these form the basic kinds of symbols. Although there is a similarity in terms to those used for geographic phenomena, it must not be assumed that point symbols are used only for point phenomena, and so on. This is far too simplistic. As will be seen, point symbols may be used to symbolize areas or volumes; lines and polygons may represent volumes, depending on several criteria.

In addition to the three basic kinds of symbol marks, there are additional *visual variables* that permit differentiation of symbols by the characteristics and values they represent. These variables were proposed by Jacques Bertin in 1983 and have since been elaborated upon by other authors. Table 7.2 illustrates the variables as conceived by different authors for static maps. Eight variables are accepted by almost all cartographers; these are form, size, hue, color value (lightness), color intensity (saturation),

---

[1]These comments refer to conventional maps; symbolization of tactile, haptic, animated, and interactive maps is discussed later.

TABLE 7.2. Comparison of Visual Variables by Cartographer

Dent et al.	Kraak & Ormeling	Krygier & Wood	MacEachren	Monmonier	Slocum et al.	Tyner
Size	Size	Size	Size	Size	Size	Size
Shape	Shape	Shape	Shape	Shape	Shape	Shape
Hue	Hue	Hue	Hue	Hue	Hue	Hue
Lightness/ value	Value	Value	Value	Value	Lightness	Lightness
Texture	Texture	Texture	Texture	Texture		Texture
Orientation	Orientation		Orientation	Orientation	Orientation	Orientation
Saturation/ intensity		Intensity	Saturation		Saturation	Saturation
Arrangement			Arrangement		Arrangement	Arrangement (pattern)
Focus			Focus			
					Spacing	
					Perspective Height	

pattern, texture, and orientation (Figure 7.5). The other variables of location, crispness, transparency, resolution, and perspective height have less acceptance.

Form or shape is the primary distinguishing variable. Differences in shape are not normally used to symbolize differences in quantity, but to identify kinds of objects—for example, circles for towns, stars for capital cities, and the like. In other words, shape differentiates between nominally classified phenomena (Figure 7.6).

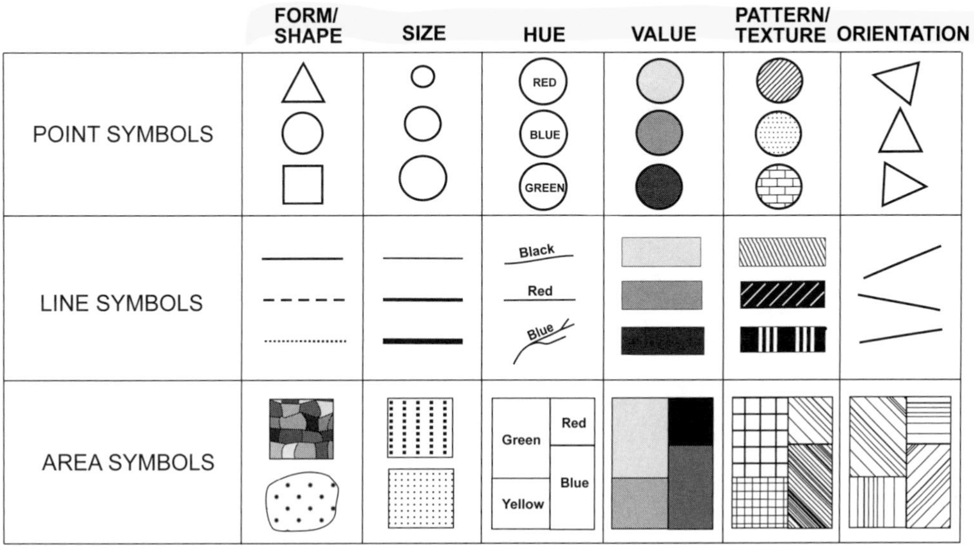

FIGURE 7.5. The visual variables of symbols.

The form may be *representational*, such as an airplane to designate airports, trees to symbolize forests, or a pick and shovel to represent a mine. These shapes have also been called *pictorial, replicative*, or *mimetic* symbols. A subgroup of pictorial symbols gives an impression of movement or action; these are called *dynamic* symbols. Examples are bomb bursts, flames, arrows, and the like. In other instances a geometric shape is used to represent types of objects; the form of the symbol has no relationship to the form of the object symbolized. Thus, a square can be used to represent an airport, a star a capital city, or a triangle a mine. These are called *abstract, arbitrary*, or *geometric* symbols. In still other instances the form of the symbol may suggest in some way the feature symbolized: a triangle for a mountain peak or regularly spaced green dots for an orchard. These have been called *associative* or *semimimetic* symbols. Although the pictorial–associative–abstract description is a useful device for visualizing symbols, there is considerable overlap between the categories, which might better be thought of as points on a continuum than as mutually exclusive categories.

Linear symbols are not usually thought of as having shape other than the contour of the feature represented. However, the form of the line symbol can be varied by making it solid, broken, dotted, and so on. Linear symbols may also be classed as pictorial, associative, and abstract.

Shape does not directly apply to area symbols because the shape of the polygon corresponds to the area represented; however, the symbol may be made up of a variety of smaller marks that themselves have shape and form patterns. In such cases, the symbol that is applied may be a tree, a dot, or a line, and these can be seen to have pictorial, associative, and abstract elements. Some books use cartograms as examples of shape, but in this book cartograms are considered as a separate type of map.

Size is normally used to show the magnitude of the data represented, that is, for ordinal, interval, and ratio representations; it is not appropriate to use size to distin-

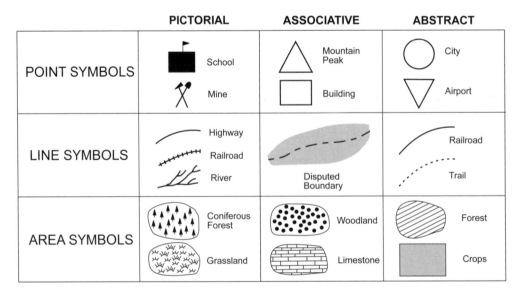

FIGURE 7.6. Shape as a visual variable.

guish between different nominal categories. If a circle is used to represent cities, the size of the circle can be varied according to the population of the city. Linear symbols may vary in width and be either single or double lines. The gauge of a line (line width) indicates relative importance or actual amounts. The size of area symbols, like shape, is governed by the extent of the area symbolized and therefore this variable does not apply in the symbolic sense.

Each of the three dimensions of color, hue, value, and intensity (see Chapter 4) can be used as a visual variable. *Hue* is a very strong distinguishing characteristic for those with normal color vision and is effective when similar symbol forms or sizes must be used, such as red lines for roads, black for railroads, blue for rivers, or different colored dots to represent different agricultural products. Normally, hues should be used to represent differences in kind, that is, different categories, rather than differences in amount. Temperature maps commonly show cool temperatures in shades of blue and warm temperatures in shades of red; here the difference between cold and hot can be conceived of as different categories.

Differences in *color value* or *lightness* refer to the range of shades from light to dark within a hue, such as light gray to black on a black-and-white map or pink to red on a color map There are limits to the number of value steps that can be used because the human eye can only distinguish six to eight steps.

Although lightness can be used to represent differences in kind within a category such as vegetation (e.g., all woodlands in lightness steps of green), it is not a good practice to use lightness to distinguish between categories. The common convention is for differences in color value or lightness to symbolize differences in degree or amount. Normally light tones are used to represent smaller amounts and dark tones are used to represent larger amounts. Differences in *saturation* combined with lightness can be used to show subcategories within a larger group, for example, vegetation types.

Symbols have not only hue or tone but also pattern or texture. *Pattern* is the spatial arrangement of smaller elements within the symbol. *Texture* is made up of an aggregate of smaller elements, such as dots that produce an overall impression of coarseness or smoothness. Pattern is less evident in fine-textured symbols than in coarse ones. Texture and pattern can only be applied to symbols that cover some area, although they are not limited to areal symbols; they can be used on circles, for example.

Orientation refers to the direction of the symbol mark or of the patterns used within the symbol. Orientation of the symbol may show the direction of the feature symbolized, such as road direction, or it may be used with pattern to represent differences in kind. If the direction of symbols has no significance to the distribution and is not a part of an areal pattern, then the symbols should be oriented or aligned with a common reference line, such as the map margins; otherwise, the reader assumes some significance for the orientation (Figure 7.7). Different orientation of patterns can be used for differences in kind, although it is not the best choice, but should not be used for quantitative maps.

Some authors consider *location* to be a visual variable. This is certainly the case when drawing roads or rivers, when point symbols are placed in their actual geographic positions, or when areas are colored according to land use. However, symbols are not always placed where the phenomena occur; sometimes this is a result of

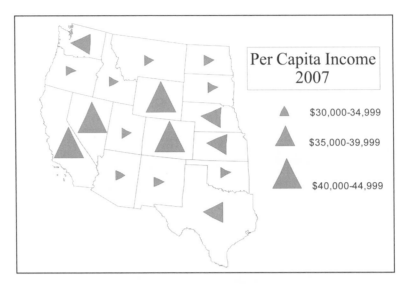

FIGURE 7.7. Varying orientation must serve a purpose. The reader would assume that the varying orientation of the triangle has some significance, but on this map it does not.

the nature of the symbol and sometimes because of the data. A line showing average traffic volume on a freeway may approximate the actual route; a circle that symbol-izes the total number of cattle for a state may be placed in the center of the state, not where the majority of cattle are found.

Boundaries, especially of areas, may not be clearly defined or there are transi-tions between one category and another. These "fuzzy boundaries" may be areas of uncertainty, or may indicate a transition; placing a solid line could mislead the reader into thinking there is a sharp boundary where there is none. In this case the variable of *crispness* can be applied. A line that appears blurry or less crisp can be used to outline the area. The variable of *transparency* can be used to indicate areas of overlap (Figure 7.8).

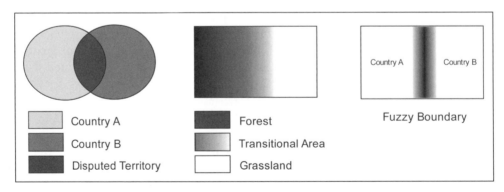

FIGURE 7.8. Crispness and transparency can be considered visual variables.

SYMBOL SELECTION AND DESIGN

Symbolizing the content of a thematic map is a complex process. No simple step-by-step guidelines or system can be provided, although there have been efforts at creating design algorithms for some map elements. Symbolization cannot be executed independently of other cartographic elements. There is no simple correspondence between dimensions of geographic phenomena and dimensions of map symbols. Although four categories of spatial phenomena exist, as we have seen, only three kinds of mark can be used to represent those categories on a two-dimensional map. Any geographic phenomenon can be interpreted in a variety of ways and can be represented with a different symbol. Figure 7.9 shows the same data with a variety of different measurements and symbolization.

The mapmaker must solve two problems in symbolization: symbol selection and symbol design. *Selection* involves determining, according to the cartographer's concept of the phenomenon, the best general type of symbol to represent it. *Design* involves consideration of the characteristics of the phenomenon, the visual variables of symbols, and the quality and level of measurement of the data to create the best symbol design for a particular purpose. Bear in mind that there is no single best way to represent the data.

General Guidelines

Although the chapters that follow present the advantages, disadvantages, and use of individual symbol types, here some general guidelines for symbol selection are given. The overriding concerns are clarity and appropriateness. Symbols must be chosen that are clear and readily identifiable from the information given in the legend. They

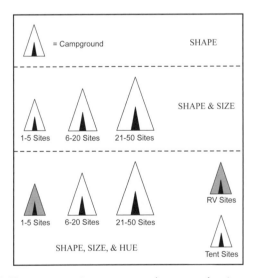

FIGURE 7.9. Several different campsite maps can be created using one basic shape. Here a simple qualitative map shows "campsites," an ordinal symbol shows number of sites, and a third variation uses tone to distinguish between tent and RV campgrounds.

must look the same on the map and in the legend. The symbol must be appropriate for the purpose and audience of the map.

Frequently, several different symbol types will initially seem equally acceptable, but upon consideration of all factors only a few will be truly suitable. Although GIS can create a variety of symbols, the ultimate choice still resides with the cartographer.

Several factors are involved in making the symbol choice: the purpose and theme of the map; the cartographer's conception of the geographic phenomena; identification of the basic spatial attributes of the phenomena; quality, nature, and measurement level of the data; scale; abilities of the map user; cartographic conventions; and compatibility of symbols.

Geographic phenomena, as we have seen, may be positional, linear, areal, or volumetric, and they may also be continuous or discontinuous. Point symbols are not limited to the representation of point data. Often it is desirable to symbolize areal data at points. Figure 7.10 shows the number of farms in the western United States by point symbols. Although the actual data apply to an entire state, they may be displayed as totals at point locations. In this example the points are at the centers of the states. Volume phenomena may be symbolized by linear or areal symbols. Data for continuous phenomena are often obtained only at certain points, for example, weather stations for temperature or rainfall or control stations for elevations. In these cases, although the data are determined at points, the actual distribution represents a volume, and the distribution is shown with a linear symbol (Figure 7.11).

The *purpose* and *theme* of the map, as we have seen, must be considered at all stages of the cartographic process but are especially important here. Several questions might be asked: What are the general and specific purposes? The general purposes of thematic maps are to provide qualitative or quantitative information for different locations or to map the characteristics of the phenomena to reveal spatial patterns

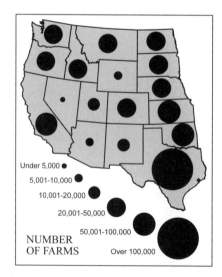

FIGURE 7.10. Point symbols can be used to aggregate data; they are not confined to one symbol representing one occurrence.

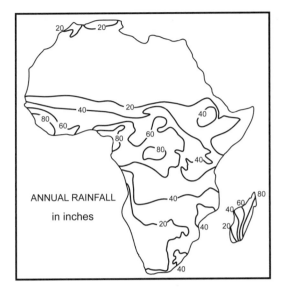

FIGURE 7.11. This map represents a volume, amount of rainfall, although the data are collected at points and represented by lines.

and emphasis. The specific purpose of the map must also be known. What exactly will the map show? What is its theme? What will it be used for?

Another vital concern is the *nature of the data.* What is their level of measurement? Are they sequential or bipolar (diverging from a central value). Lower levels of information cannot be portrayed by higher levels of symbolism. That is, data that are in ranked form cannot be symbolized on an interval scale. If it is desirable for the reader to be able to make estimates of value, the data must be collected with this end in mind. The quality of the data is reflected in the quality of the map. A good map cannot be made with poor data.

In addition, the visual variables do not work equally well for all types of data. As we have seen, form works best with qualitative data and lightness works best with quantitative data. Table 8.4 shows visual variables and the data types with which they work best.

Map users vary widely in their map-reading abilities. Map interpretation is a skill that must be learned; it is not inborn, although some people seem to grasp the concepts more quickly than others. Since maps are, at least implicitly, aimed at a specific type of user, the symbolism chosen should suit the presumed abilities of that user. Usually, the cartographer does not have a personal acquaintance with the proposed user or group of users, so knowledge of these abilities is inferred from place of publication. Even among users with roughly equivalent abilities, perceptions must be considered. A major area of cartographic research has been the perceptual abilities of users. Many studies were carried out in the 1970s and 1980s on how readers perceive symbols and how well symbols communicate information. Maps made for the visually impaired have different requirements than those prepared for readers with normal sight. Since the advent of the computer, animations and interactive maps have been studied.

Certain *conventions* in symbols are firmly established. An especially strong convention is the use of blue for water, although some water features actually appear green or brown. There is no rule, other than common usage, that requires these conventions to be followed, but readers sometimes object or are confused if the conventions are not observed. The cartographer should follow established conventions unless there is good reason to depart from them. Cartographic design and symbolism would stagnate without innovation, but the meaning of symbols should always be clear, especially if unconventional marks are used. Conventions will be discussed in the context of specific symbol types.

If more than one type of symbol is used on a map to distinguish between categories, to illustrate a relationship between two or more distributions, or merely to display base map and thematic information, it is important that the symbols chosen be *compatible*. They must be clearly distinguished from one another, they must not interfere with the recognition of the complete pattern, and they must not overlap one another to the point of elimination. The symbols must also be compatible with any background colors or tones or they will be undecipherable (Figure 7.12).

For the meaning of symbols to be clear and unambiguous for the reader, the symbols must be identified in a legend. Any symbol used on the map must be explained in the legend unless its meaning is so obvious that no confusion could result. As noted in Chapter 2, the appearance of the symbol on the map and in the legend must be identical or the map will be confusing to the reader.

The next chapter treats the symbolization of various types of geographical phenomena and provides more detailed guidelines on the selection, use, and design of specific symbols.

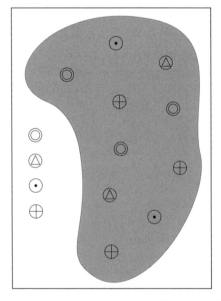

FIGURE 7.12. Symbols must be compatible with the background. Here it is difficult to distinguish the symbols against the dark tone.

SUGGESTIONS FOR FURTHER READING

Bertin, Jacques. (1983). *Semiology of Graphics* (William J. Berg, Trans.). Madison: University of Wisconsin Press.

MacEachren, Alan M. (1994). *Some Truth with Maps: A Primer on Symbolization and Design*. Washington, DC: Association of American Geographers.

MacEachren, Alan M. (2004). *How Maps Work: Representation, Visualization, and Design*. New York: Guilford Press.

Tufte, Edward R. (1983). *The Visual Display of Quantitative Information*. Cheshire, CT: Graphics Press.

Tufte, Edward R. (1990). *Envisioning Information*. Cheshire, CT: Graphics Press.

Chapter 8

Symbolizing Geographic Data

> Excellence in statistical graphics consists of complex ideas
> communicated with clarity, precision, and efficiency.
>
> —EDWARD R. TUFTE, *The Visual*
> *Display of Quantitative*
> *Information* (1983)

As we saw in the previous chapter, visual variables are used to determine symbol types to represent geographic phenomena and to vary their design. However, a number of kinds of symbols are used on maps. I will discuss symbolizing data at points, along lines, over areas, and as volumes.

SYMBOLIZING DATA AT POINTS

Two kinds of data are symbolized at points: those that actually occur at points, such as locations of places on small-scale maps, and data that are aggregated at a point, such as totals for counties or other enumeration areas. In one sense, this is a function of scale. Cities, for example, actually cover area, but even a sprawling city like Los Angeles is merely a point on a small-scale world map.

Both qualitative and quantitative data may be shown by point symbols. *Qualitative* symbols may be pictorial, associative, or geometric and most often use the variables of shape and hue. Qualitative maps tend to be fairly straightforward. *Quantitative* point symbols may also be pictorial, associative, or geometric and most often use the variables of size and location (Figure 8.1).

Qualitative Point Symbols

The variables used will be shape, hue, or pattern, and the symbol is usually placed in its correct geographic location. Qualitative point symbols should not vary in size or

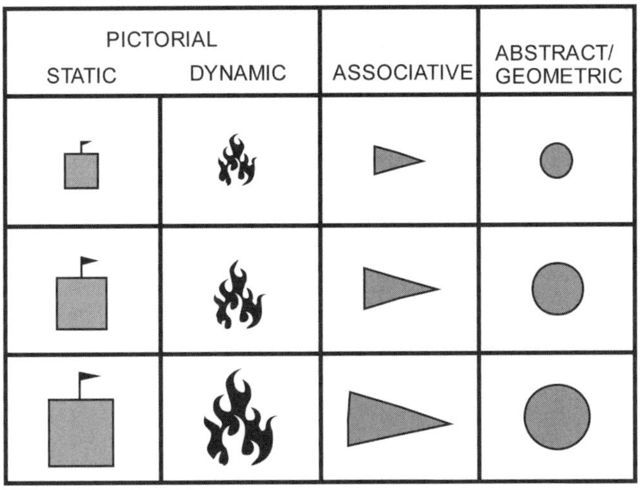

| | PICTORIAL | | ASSOCIATIVE | ABSTRACT/ GEOMETRIC |
STATIC	DYNAMIC			

FIGURE 8.1. Quantitative point symbols can be pictorial or abstract and the variable of size represents the amount.

tonal value since these variables are used to symbolize differences in size or amount. A particular concern is legibility, especially when a variety of shapes are used. Although symbol sizes will not vary on the map, the size and shape must be large enough to distinguish the individual symbols. This can be a problem especially with maps viewed on monitors–if the symbols are too small it is difficult to differentiate between squares and circles. Complex pictorial symbols with letters within shapes can fill in if they are reduced.

Most illustration and GIS software have a variety of pictorial or abstract point symbols available.

Quantitative Point Symbols

Quantitative point symbols are of two basic types, dots and proportional figures, but there is variation within these categories.

Dot Maps

In the most basic form, dot maps use a single dot ("dots" may actually be circles, squares, triangles, or other shapes) to represent a given quantity, such as one dot represents one house, or one dot represents 500 people. The dot is placed at the location of the phenomena for single units or in the center of the distribution for multiple units. The purpose of dot maps is to show the distribution of the phenomena; they are not created to determine quantities, that is, the individual dots cannot be counted to find totals. These maps are called *simple dot maps* or sometimes *dasymetric dot maps* (Figure 8.2). Because the dots are placed in the area of greatest density, the map shows areas of concentration and sparseness. They can be very effective maps.

Making a simple dot map requires two kinds of data: quantity and location. Quantities can be determined from censuses or other tallies, and location can be

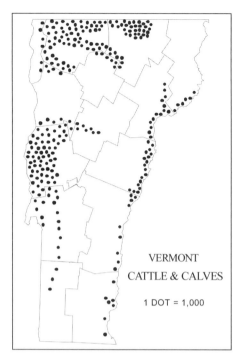

VERMONT
CATTLE & CALVES
1 DOT = 1,000

FIGURE 8.2. A simple dot map uses one dot to represent a given quantity and to show density and distribution. This map has a good balance of dot value and size.

determined from related maps, fieldwork, and/or from imagery. Related maps might show where a crop cannot be grown because of soil or slope. Imagery can show the actual crop in place.

While the size of individual dots doesn't vary, the choice of dot size and value is important. A dot that is too small is hard to see and can be mistaken for a flaw in the paper or an errant pixel on a monitor; dots that are too large stack on top of one another and obscure patterns. Likewise, if the value of the dot is too large, patterns can't be seen and if it is too small, areas appear too dense (Figure 8.3).

In some cases, two or more phenomena may be represented on the same map, such as spring wheat and winter wheat or cattle and sheep. In these instances, different colored or shaped dots may be used to distinguish the phenomena.

Ironically, the simple dot map, which was commonly used until the advent of GIS and was one of the easiest maps to construct, is a difficult map to create with a computer. There is, at this writing, no software to create such maps although they can be constructed by using imagery as one layer, outlining the areas of occurrence, and placing the individual dots on another layer. This is a cumbersome task and not often done. A valuable map type has fallen into disuse as a result.

A more commonly used type of dot map now is the *dot density map* (Figure 8.4). On such maps, the dots are placed randomly within the enumeration area. If the enumeration area is large, such as a state, these maps are somewhat crude and can be misleading; if the enumeration area is small, such as census tracts or blocks, the resulting map approaches the quality of a simple dot map. If the data are available

Los Angeles Long Beach	Size	S i e r r a N e v a d a Signal Hill	Extent (letter spaced, normal)
Land *Water*	Form (Roman, Italic)	**Los Angeles** Long Beach	Extent (normal, condensed)
Los Angeles **Long Beach**	Form (type face)	Los Angeles Long Beach	Hue
Los Angeles Long Beach	Intensity (bold, normal)	Los Angeles Long Beach	Value
Los Angeles Long Beach	Value		

PLATE 3.1. The visual variables of type.

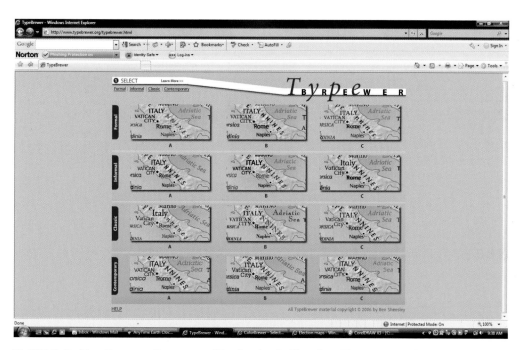

PLATE 3.2. TypeBrewer. Courtesy of Benjamin C. Sheesley, *TypeBrewer.com*

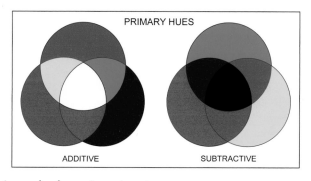

PLATE 4.1. Additive and subtractive primaries.

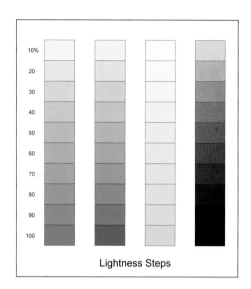

PLATE 4.2. Lightness steps of the subtractive primaries and a neutral scale.

Saturation

PLATE 4.3. Saturation steps.

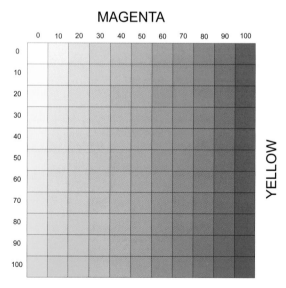

PLATE 4.4. The range of colors possible using just two subtractive primaries.

PLATE 4.5. Adding color to the route map in Figure 4.6 enhances its readability.

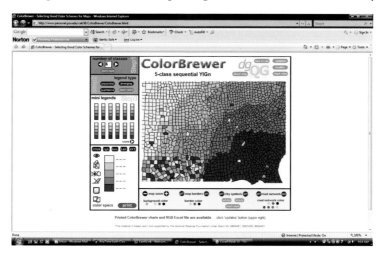

PLATE 4.6. ColorBrewer. Courtesy of Cynthia Brewer, *ColorBrewer.org*

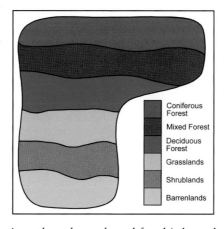

PLATE 4.7. There is no logic to the colors selected for this hypothetical vegetation map.

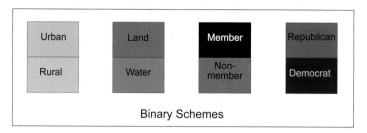

PLATE 4.8. Binary schemes.

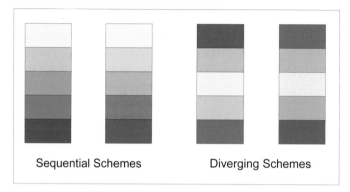

PLATE 4.9. Sequential and diverging schemes.

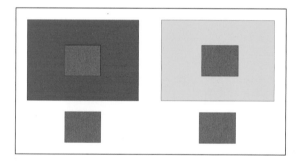

PLATE 4.10. The gray squares are all the same shade of gray.

PLATE 4.11. Care must be used when choosing colors for type that will be on a colored background.

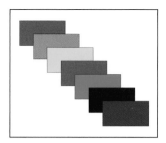

PLATE 4.12. These colors are reproduced in black and white in Figure 4.10.

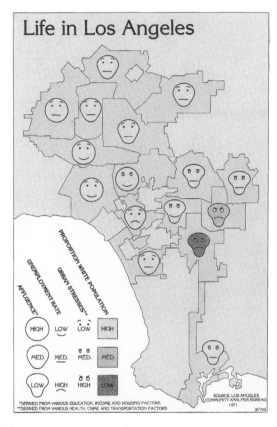

PLATE 9.1. Chernoff face map. Courtesy of Eugene Turner.

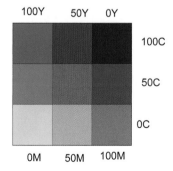

PLATE 9.2. Color chart for the bivariate choropleth map in Plate 9.3. The subtractive primaries are used.

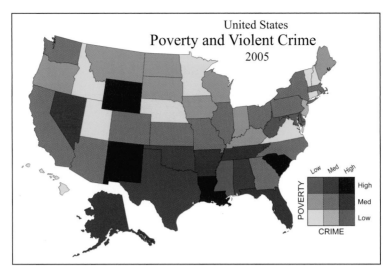

PLATE 9.3. Bivariate choropleth map combining data from the maps in Figure 9.1.

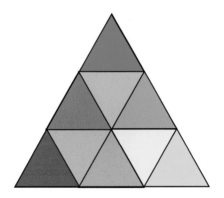

PLATE 9.4. A simple trivariate chart.

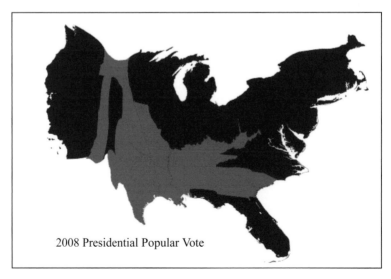

PLATE 10.1. Map showing the popular vote of the 2008 presidential election. Courtesy of Mark Newman, University of Virginia.

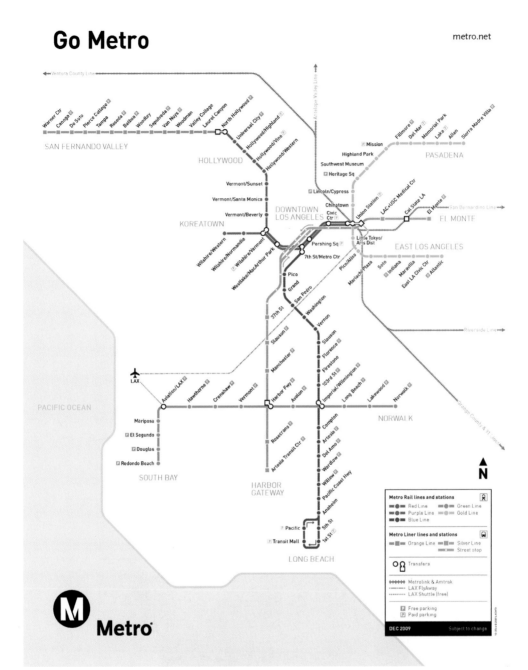

PLATE 10.2. Transportation diagram for Los Angeles Metro. Courtesy of Metro © 2009 LACMTA.

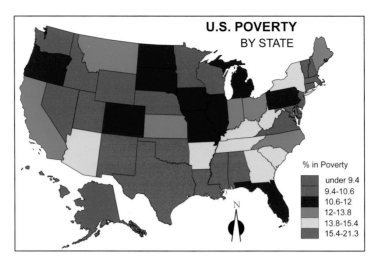

PLATE 12.1. Magazine map "before."

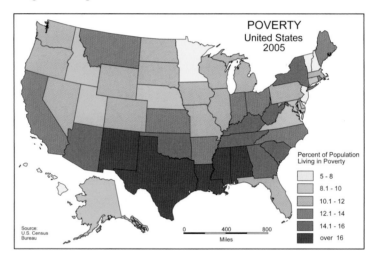

PLATE 12.2. Magazine map "after."

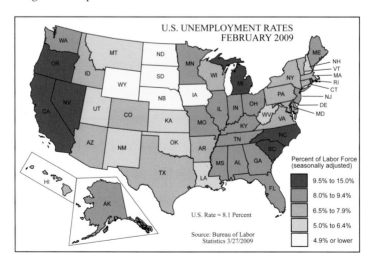

PLATE 12.3. Newspaper map "after."

only for large areas, and density isn't a primary concern, it is better to choose a different kind of symbol, such as a proportional figure or choropleth.

Proportional and Range-Graded Point Symbols

Symbols of varying sizes can be used to symbolize totals at a point. The shape most frequently used is a circle, but squares, triangles, and even pictorial shapes can be

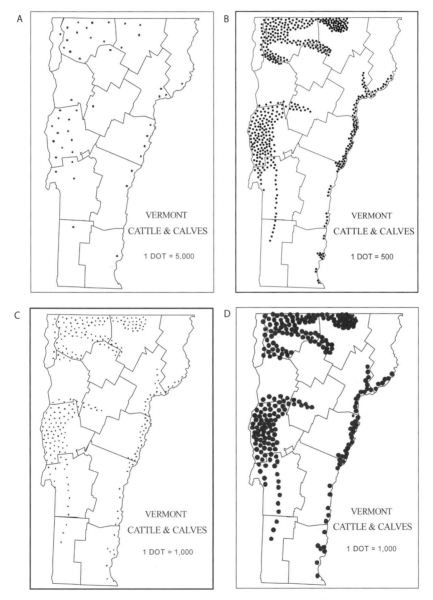

FIGURE 8.3. (A) The dot value is too large so the pattern can't be seen; (B) the dot value is too small, which also obscures the pattern; (C) the dot value here is good, but the dots are too small to be easily seen; (D) the dots here are so large that they overwhelm the map.

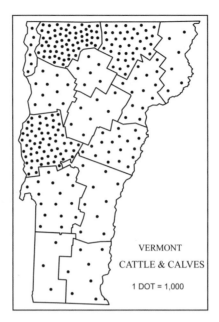

FIGURE 8.4. Dot density maps have dots placed randomly within the enumeration areas.

used. The size of the figure represents the quantity. *Proportional* figures are constructed so that the *area* of the figure is proportional to the value: $A = \pi R^2$. Thus, a circle representing 1,000 people will be one-half as large in area as one representing 2,000 people. It is important to recognize that it is the area of the circle, not the radius, that is two times as big (Figure 8.5). In this example, the radius of the larger

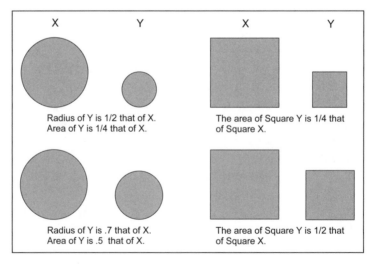

FIGURE 8.5. For proportional figures it is the *area* of the figure that is important. If the circle radius of the large circle is two times that of the small circle, then the small circle is only 1/4 the area of the larger circle.

circle would be 1.4 times larger than the smaller circle. One advantage of creating proportional symbols with computer software is eliminating the drudgery of calculating radii.

Graduated and proportional symbols are used for the following:

1. *When the range of data is too great to be represented by dots.* If the dots in the most dense area stack on top of one another so that the enumeration area appears solid black, a proportional symbol might be a better choice.

2. *To symbolize totals over an area,* such as total water power for a region.

3. *To symbolize totals at a point,* such as populations of cities. Sometimes these symbols are combined with dots, with dots showing rural population and circles showing urban population.

4. *To compare one subdivision with another,* for example, counties within a state or states or provinces within a country.

For proportional symbols, a different size can be calculated for each enumeration area; thus for county populations in California, fifty-eight circles would be used. Figure 8.6 is a proportional circle map for the San Francisco Bay Area. It and the variations in Figures 8.7, 8.8, and 8.9 were created from the data in Table 8.1.

A problem with truly proportional circles is that readers are not likely to calculate the value for each of the circles; indeed, this information can be shown more effectively in a table. Thus, the range of data can be divided into several categories and those categories are represented in the legend. The figures are drawn proportional to the midpoint of the range for each category. These are called *range-graded* figures (Figure 8.7).

A third way of representing data with figures of different sizes is simply choosing

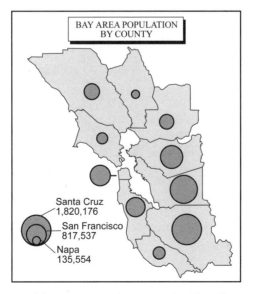

FIGURE 8.6. The circles on this map are truly proportional to the quantities they represent.

TABLE 8.1. Bay Area Population

County	Total population
Alameda	1,530,620
Contra Costa	1,044,201
Marin	256,310
Napa	135,554
San Mateo	734,453
San Francisco	817,537
Santa Clara	1,820,176
Santa Cruz	265,183
Solano	423,970
Sonoma	482,034

a series of arbitrary sizes (graduated symbols) that are not proportional. If the purpose is merely to show quantities, not relate specific values to one another, this can be a simple method. This also works for data that are ranked or ordered, but have no values attached, such as small, medium, large or village, town, city, metropolis (Figure 8.8). It is important that these circles be shown clearly in the legend.

Other proportional figures such as squares, triangles, and even human figures can be used, but not as much research has been undertaken on these as on circles. The principle is the same as for graduated circles—that is, the area of the figure is proportional to the quantity represented. The size of squares may be easier to estimate for readers, and it may be easier to compare areas of squares than areas of circles (Figure 8.9). There are, however, some design problems that arise for these symbols. It is more difficult to orient squares than circles on the map if the projection used has radi-

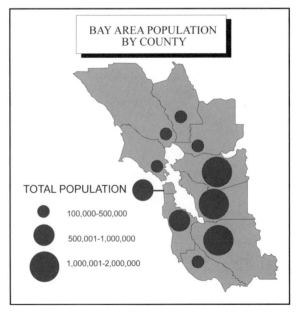

FIGURE 8.7. Range-graded circles. Here the circles are drawn proportional to the middle of the category ranges.

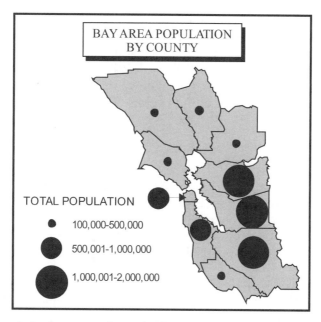

FIGURE 8.8. The circles here are arbitrarily chosen and are not proportional to the quantities.

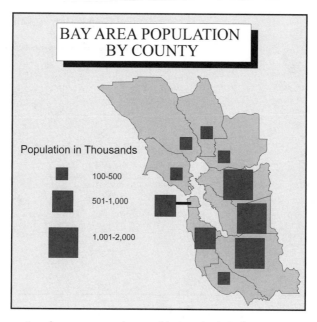

FIGURE 8.9. Proportional squares can be used instead of circles.

ating meridians. The squares cannot be oriented to the meridians without creating a "tipsy" appearance; therefore they must be lined up with the borders of the map. The same problems are encountered with triangles and other geometric figures.

Segmented proportional symbols or *pie charts* are used in combination with proportional or point symbols to show a breakdown of total value, such as independent voters as a percent or proportion of total voters. These will be described in Chapter 9 on multivariate data.

Proportional Spheres and Cubes

If the range of data is so great that the largest circle overwhelms the map and the smallest figures are insignificant (Figure 8.10), a three-dimensional figure such as a sphere, a pyramid, or a cube can be used. For such symbols, the *volume* of the symbol, not the area, is proportional to the quantity represented. These figures can result in very attractive maps, but readers have difficulty in interpreting some of these symbols.

Proportional spheres are probably the most difficult to interpret. First, the same problem arises as with circles, that is, estimating and comparing sizes. Second, because the symbols only simulate the third dimension and are placed on a two-dimensional map, readers have a tendency to compare or estimate areas rather than volumes. The procedure of creating proportional spheres is similar to that for creating graduated circles except that the cube root of the quantity is used instead of the square root because the volume of a sphere is represented by the formula $V = \pi R^3$.

Columns or piles of blocks whose volumes are proportional to the amount represented are also used. While these are striking, some users have difficulty visualizing

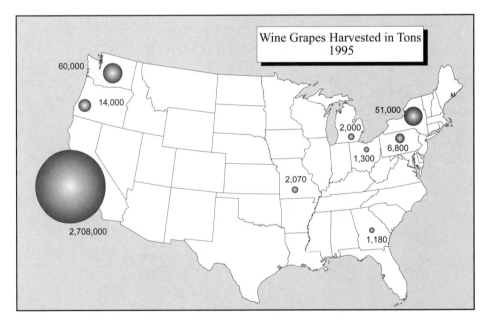

FIGURE 8.10. Proportional spheres are used if the quantities have a wide range of values. The *volume* of the figure is the value.

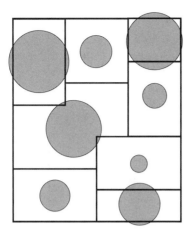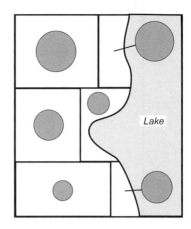

FIGURE 8.11. Ideally, circles are not larger than their enumeration areas; if some are, callouts can be used.

the number of blocks since some are necessarily hidden. Little has been done in testing reader perceptions of these symbols.

Design Considerations

Placement of Circles

If circles represent point phenomena, they are ideally located at the point. For circles that represent aggregates over regions, the options are to place the circle in the center of the region or over the most dense area. In most cases it is best to center the circles within the enumeration areas. Ideally, the circle fits entirely within the enumeration area and doesn't overlap boundaries; this is a major factor in determining the radius of the circle. However, because large quantities are often found in small areas, this is not always possible. In such cases, it may be necessary to place a circle outside the enumeration area with a callout, as in Figure 8.11.

Filling Circles

Often, proportional symbols are filled with a hue or pattern to make them more legible because an open circle lacks contrast. Unless the circle represents two variables it is not good practice to use lightness as well as hue, such as shades of gray or red, in an effort to enhance the quantities represented. This is called redundancy and although there is some debate over the practice, there has been no testing with map users. Redundancy can be confusing to the reader who assumes that the lightness steps represent a second variable (Figure 8.12).

Overlapping Circles

Frequently, circles overlap, as in Figure 8.13, where a number of cities are quite close to one another. If the circles aren't filled with a hue or pattern, they appear as a tan-

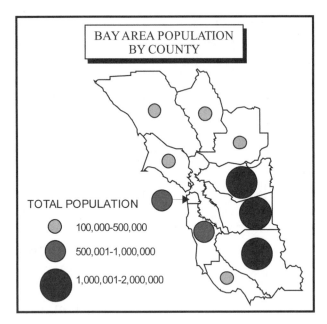

FIGURE 8.12. Redundancy with lightness and size. Some cartographers feel this reinforces the information, others believe it is confusing.

gled mass; if they are colored a solid hue, individual circles are difficult to distinguish and may be "blotted out" entirely. Solutions to this problem are *haloing* and using transparent circles. Haloing involves surrounding the circles with white or a lighter tone. Transparent circles appear to be made of translucent or transparent layers so that all of the circles can be seen. It is possible to draw circles that have a fill color, gray tone, or pattern, but no outline. This tends to be a matter of personal preference; it is not a subject that has been given much attention. To some, however, the lack of an outline looks unfinished. Color and tonal choices are made according to the guidelines in Chapter 4.

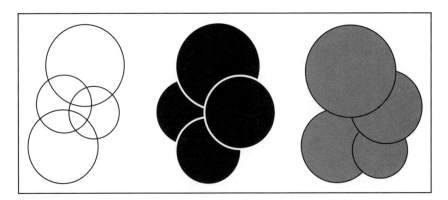

FIGURE 8.13. When circles overlap they can be confusing; filling with masks and halos can make the information more clear.

Legends for Proportional and Graduated Symbols

In part the decision for type of legend is determined by the amount of space available, but it is also governed by the type of proportional figure and the capabilities of the software. Choices range from a simple statement to a complete range of circle sizes. Figure 8.14 shows a variety of legend types. The nested type, Figure 8.14a, conserves space; the style in Figure 8.14b allows the reader to estimate sizes for intermediate circles not shown on the legend; and the style in Figure 8.14c, while using more space, allows display of all circles for a range-graded map.

SYMBOLIZING DATA ALONG LINES

As with point data, both qualitative and quantitative data may be shown with lines.

Qualitative Line Data

Qualitative line data are symbolized with hue, form, pattern, and orientation. A common example is a transportation map that uses different hues for roads, railroads, and rivers. Maps of explorers' routes can distinguish different explorers with different hues and patterns for routes, as seen in Figure 4.6 and Plate 4.5. Different line widths imply a difference in size or ranking, so they should not be used unless such a ranking or hierarchy is present, as with political boundaries.

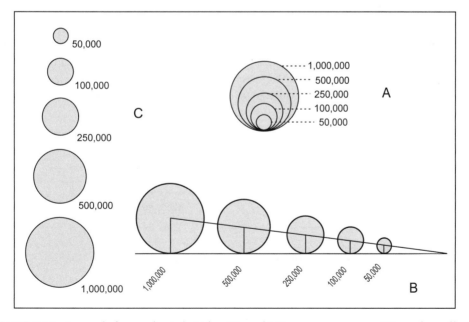

FIGURE 8.14. Legends for graduated circles. Nested (A), conserves space; (B) works well with truly proportional circles because it allows the reader to estimate circle sizes; (C) can be used for any type.

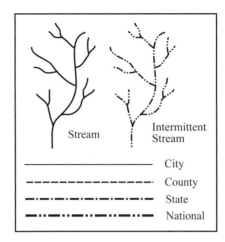

FIGURE 8.15. Quantitative line patterns. These show hierarchies and are used for ordinal rankings.

Quantitative Line Data

If the data are in ordered or ranked categories that imply differences in size, line width can be varied. Thus, freeways are larger than state highways which in turn are larger than county roads. In this case, the different categories may also be shown in different hues as well. Boundaries are usually hierarchical so the lines for national, state, county, and minor civil divisions should reflect that hierarchy with variations in width and/or lightness (Figure 8.15).

Often data are collected that show amounts along routes. These can be shown with lines that vary in width, called *flow lines* (Figure 8.16). Examples are number of cars along a road, amounts of a product shipped between two points, and number of immigrants moving between two locations. For such symbols, the width of the line is drawn proportional to the amount represented. The line may follow the actual route, such as number of cars, or the route may be schematic, as for shipping or immigration.

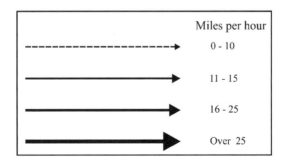

FIGURE 8.16. Flow lines can be ordinal or interval.

SYMBOLIZING AREAL DATA

Representing qualitative data with polygons is straightforward. The variables most often employed are hue, pattern, and orientation. Different hues represent different nominal categories, such as vegetation types, or land-use categories on color presentations. The main concern here is that there be logic to the choice of hues, as we saw in Chapter 4.

If color is not an option, the mapmaker must rely on pattern for different categories. Shades of gray cannot be used because these imply differences in amount. GIS and illustration software have a number of stock patterns available and some permit the user to design his or her own. These patterns may be pictorial or abstract, made up of varying line patterns and orientations.

Representing Uncertainty and Overlap

Phenomena in the real world are not always clear-cut even though they appear so on many maps. There are many instances of uncertainty, overlap, and even simultaneous occupation of phenomena. Two examples are language and religion. One method that has been used is interdigitation, which shows a blended area; a second is overlapping patterns that form a third pattern; or a third is transparent colors that blend into one another (see Figure 7.8). Lines at boundaries can also be portrayed as fuzzy or blurred. The method used depends on the capabilities of the software used.

SYMBOLIZING VOLUME DATA

Many geographic phenomena can be thought of as volumes. If a phenomenon occurs over an area rather than at a point or along a line, and the data have magnitude, then the data values may be thought of as having height and the phenomenon as three-dimensional. For example, land surface can be visualized as a volume whose three dimensions are latitude, longitude, and elevation above sea level for each point.

For other kinds of data, such as population, rainfall, or housing values, the volume cannot be seen, but we can conceive of a three-dimensional surface with hills and valleys of population, rainfall, or housing values. This imaginary three-dimensional surface is called a *statistical surface* (Figure 8.17). The concept of a statistical surface is especially useful in visualizing the nature of statistical distributions and in symbolizing those distributions. If the coordinate system of the statistical surface is represented graphically, the locational coordinates are on the x and y axes, and the value is represented along the z axis and is often called the z-value (Figure 8.18).

If a phenomenon is found everywhere within the mapping area, such as temperature, that is, it is continuous, the surface is smooth and undulating. If there are sharp breaks or areas with an absence of the phenomenon, that is, discontinuous, such as population, the surface is step-like, made up of a series of cliffs and plateaus.

The statistical surface may be symbolized with either area or line symbols, depending on the nature of the surface, the way in which the data are gathered, and the aspect of the surface that is of greatest interest.

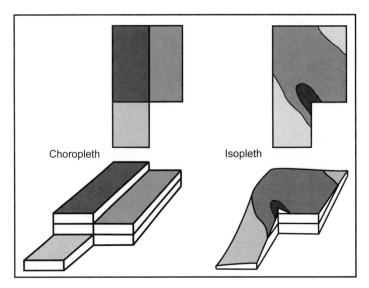

FIGURE 8.17. The statistical surface formed by choropleth and isopleth symbols. Adapted from Thrower, Norman J. W. (2008). Copyright 2008 by University of Chicago Press. Adapted by permission.

Symbolizing Volumes with Polygons: Choroplethic Maps

If data have been collected for statistical areas, such as census tracts, counties, and the like, instead of at sample points, such as temperature or rainfall, or if the distribution is discontinuous, the statistical surface will be step-like, and one of the choroplethic techniques will be a suitable method for symbolizing the data.

On a *choropleth* map, each enumeration area is shaded or colored uniformly according to the value represented. There are three kinds of choroplethic maps: the simple choropleth, the dasymetric choropleth, and the unclassed choropleth. The visual variables most appropriate are value and texture. Choropleth maps are the most common type of quantitative map produced with GIS.

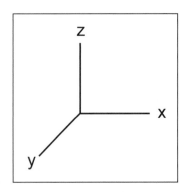

FIGURE 8.18. The statistical surface is an imaginary 3D surface where x and y represent location and z represents value.

Simple Choropleth

Conceptually, the simple choropleth is a relatively uncomplicated type of symbol. Values are determined for each enumeration area and then placed into categories or classes; shading patterns or hue lightness steps are chosen for each category and applied to the enumeration areas. The number of classes is limited by the number that the human eye can recognize. Very little information is needed other than boundaries and values (Figure 8.19).

LIMITATIONS OF SIMPLE CHOROPLETH

Because shading or patterns cover each enumeration area uniformly, it is not possible to show variation within enumeration areas. This is a limitation of the choropleth symbol. Thus, a choropleth map should not be used when the purpose is to compare enumeration areas. Figure 8.20 illustrates this problem. In the county highlighted, the population is concentrated in one corner, but the uniform shading can give an inexperienced reader the impression that population is evenly spread throughout the county.

The boundary lines on choropleth maps have no values attached to them; they are simply the lines outlining the enumeration area polygon: county or state boundaries, census tract borders, or Zip code boundaries, for example. A change in pattern or color does not represent a change in value along the boundary line. It only tells the reader that the two adjoining *areas* have different values. This is a second limitation of choropleth maps: exact values cannot be determined.

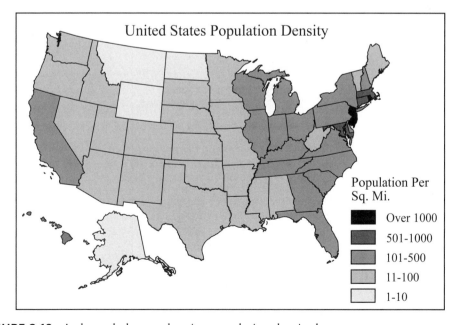

FIGURE 8.19. A choropleth map showing population density by state.

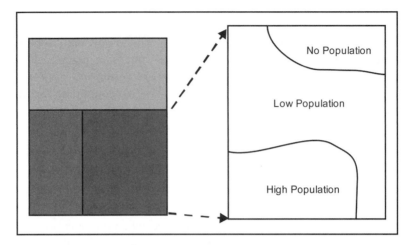

FIGURE 8.20. Choropleth maps cannot show variation within the enumeration area.

DATA CONSIDERATIONS

Unless the enumeration areas are all the same size, densities, ratios, percentages, or other derived values should be used, rather than absolute values, for the map to be meaningful. Figure 8.21 shows two areas, each of which contains 5,000 people. If the two areas are shaded according to the total population, both will be shaded alike. It is obvious that the density of population is greater in the smaller area and that the map is misleading when the population characteristics appear equal. If the two areas are shaded according to population per square mile, area B will show 200 people per square mile and area A only 50 people per square mile. This becomes especially

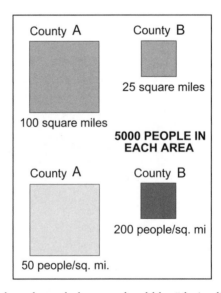

FIGURE 8.21. Values used on choropleth maps should be "derived" (e.g., ratios, percentages, or densities), or the map is misleading.

important when mapping county or state data for the United States where such areas vary widely; Rhode Island has 1,545 square miles, while Alaska has 663,267 square miles. San Bernardino County in California contains 20,106.4 square miles, which is larger than many states, while San Francisco County contains only 231.9 square miles. The map in Figure 8.22 uses raw data for population, which is not as meaningful as a map of population density.

Dasymetric Choropleth Technique

The *dasymetric choropleth* technique, although not widely used, was devised to overcome some of the limitations of the simple choropleth. Specifically, it takes into account variations within enumeration areas and therefore permits a somewhat better graphic view of distributions. Like the simple choropleth, dasymetric areas are shaded uniformly, and categories, which are usually derived values, are used. Unlike the simple version, however, areas of similarity and zones of rapid change are determined. Figure 8.23 illustrates the procedure. First, areas are determined and densities or ratios computed. The lines that separate areas have no value; they mark enumeration boundaries and zones of rapid change between categories.

This technique has some of the limitations of the simple choropleth. Exact values cannot be determined within the area from the map, and like the simple choropleth it is not designed to represent smooth, continuous surfaces. However, since the areas chosen have inherent homogeneity, dasymetric maps can come closer to reality than simple choropleths.

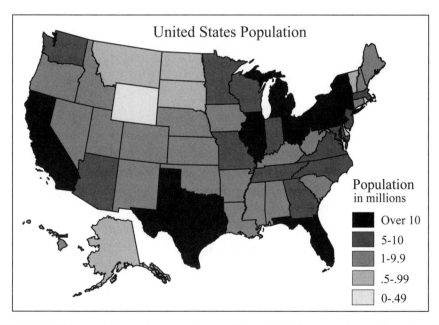

FIGURE 8.22. This map shows the same data as Figure 8.19, but raw data rather than derived values are used. Notice the difference in New Jersey and Rhode Island, both high-density states, and California and Texas, both large states with high populations but medium to low densities.

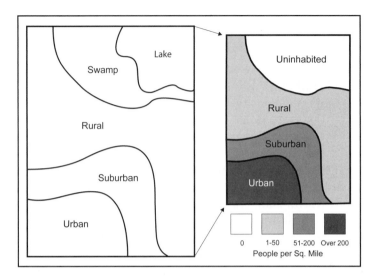

FIGURE 8.23. Creating a dasymetric map.

One reason that the dasymetric choropleth is used less than the simple choropleth is that the technique requires more information than what appears for an enumeration unit in the census. Local knowledge of the area is necessary, which may be obtained through fieldwork or imagery, to see where the subject data are concentrated.

Unclassed Choropleth

The *unclassed* or *classless choropleth* is a symbol suggested by Waldo Tobler in 1973. Although not difficult in concept, this technique was not feasible before the era of computer-assisted cartography because of the great difficulty in creating the areal symbols for enumeration areas. Cross-hatch patterns that vary in density were first used to represent each enumeration area value, that is, each value on the map has a separate pattern density. Now a continuous range of grays is used to create such maps. Essentially no generalization is done for the reader. The unclassed choropleth is the most exact representation of the data model possible, but it does not permit the cartographer to stress specific characteristics of the distribution or present a regionalization (Figure 8.24). It is useful in seeing the overall picture as a stage in classification for simple choropleths.

Design Considerations for Choropleth Maps

Because choropleth maps use area symbols and represent quantities, the variables most appropriate are color lightness, color saturation, and pattern combined with texture. Pattern isn't the first choice, but simple line or cross-hatch patterns combined with varying textures were the only option for early computer-drawn maps. Color lightness alone only permits a limited number of steps (also called *ramping*). The limitation is owing to the number of steps that the human eye can distinguish. The range can be extended by adding saturation (see Chapter 4).

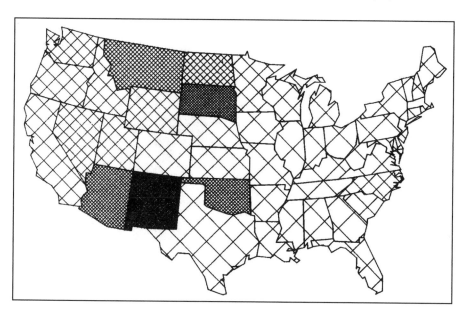

FIGURE 8.24. A computer-generated "unclassed choropleth" map. Each state has a different pattern.

Another consideration in choosing colors is the nature of the data. As we have seen in Chapter 4, a variety of color schemes are possible and one should match sequential data with sequential schemes and double-ended (diverging) data with diverging schemes.

Legends for choropleth maps normally show the various categories and the values associated. Since the lines have no value, it is best to put the category values next to the pattern or shade representing the value, as in Figure 8.25. There is some disagreement over whether the highest values should be placed at the top or the bottom of the legend, but as long as the legend is clear, high to low or low to high is a matter of preference.

Data Analysis for Choropleth Maps

A variety of techniques both mathematical and graphical can be used to analyze the data and aid in choosing an appropriate series of categories. Although category types can be selected from the software, the methods of selection are described here to aid in understanding the categories and choosing appropriate categories for a map.

EXAMINING THE DATA

A first step in the analysis of data is to place them in rank order on a spreadsheet. That is, the data are sorted from highest to lowest. An inspection of the ordered data can often give an idea of the nature of the distribution, but creating a histogram or number line will graphically show its nature. A *histogram* (Figure 8.26) is a type of bar graph for which the area of the bars is proportional to the frequency of the obser-

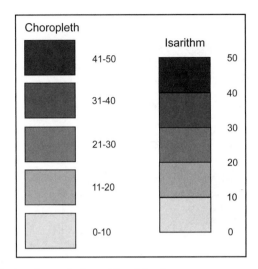

FIGURE 8.25. Legends for choropleth and isarithmic maps.

vation. The histogram can be examined for obvious breaks in the distribution that are used as a basis for the natural breaks method or to determine if the distribution is normal, equally frequent, or resembles an arithmetic or geometric progression. A *number line* is a line on which all places are plotted as points (Figure 8.27). It can represent data values, natural breaks, and divisions between classes.

An unclassed choropleth (Figure 8.24) can be used to see the spatial arrangement of the data and aid in choosing categories; however, most often categories are

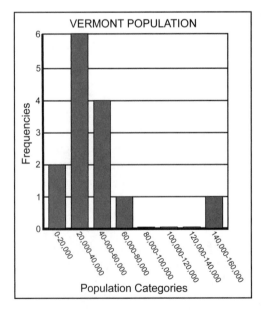

FIGURE 8.26. A histogram can be used to aid in determining class intervals.

FIGURE 8.27. A number line.

determined by examining the data. With computerized mapping, the categories can be experimented with easily.

Category Types

In general there are several category types that may be used for choropleth maps: equal steps, standard deviations, quantiles, arithmetic progression, geometric progression, and natural breaks. These are described here briefly, but since a large part of cartography and GIS involves spatial analysis, it is strongly recommended that GIS and cartography practitioners acquire a background in statistics. An advantage of computer methods is that one can experiment with different categories to visualize the distribution. As can be seen from the following discussion and the accompanying figures, the resulting maps vary in appearance and information depending on the method of categorization. The maps have been constructed from the data in Table 8.2.

Equal steps are constant intervals. They are found by determining the range of the data; the range is then divided by the number of classes, which gives the common difference. The class limits are established by starting at the lowest value and adding the common difference to get the upper limit of the first class, adding the common difference to this to get the limit of the second class, and so on, until the upper limit of the data is reached (Figure 8.28).

Standard deviations are a second type of constant interval. This method is most effective if the distribution approximates a normal distribution (bell-shaped curve),

TABLE 8.2. Arizona Population Density

Arizona	45.2
Apache	6.2
Cochise	19.1
Coconino	6.2
Gila	10.8
Graham	7.2
Greenlee	4.6
La Paz	4.4
Maricopa	333.8
Mohave	11.6
Navajo	9.8
Pima	91.9
Pinal	33.5
Santa Cruz	31.0
Yavapai	20.6
Yuma	29.0

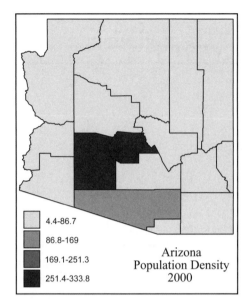

FIGURE 8.28. Choropleth map using equal steps.

which is rare for geographic distributions. Standard deviation is a measure of dispersion; it is a way of stating how far the observations vary from the mean. It is defined as the square root of the arithmetic mean of the squared deviations from the mean. The boundaries of the class intervals are found by adding or subtracting the standard deviation.

Quantiles are categories that contain an equal number of enumeration areas (observations) in each (Figure 8.29). If the areas involved vary widely in size, quantiles may be misleading.

An *arithmetic progression* is a sequence of numbers in which each term after the first is determined from the preceding one by adding to it a fixed number called the *common difference*, for example 2; 2 + 1,000; 1,002 + 1,000; 2,002 + 1,000; and so on. The common difference here is 1,000. If the distribution graph approximates that of an arithmetic progression, then it is possible to use this as the basis for categories (Figure 8.30).

Geometric progression is a sequence of numbers in which each term after the first is determined by multiplying the preceding term by a fixed number called a *common ratio* (Figure 8.31). If 2 is again the first term and the ratio is 10, the progression is 2, 2(10), 2(10)^2, 2(10)^3, or 2, 20, 200, 2,000, and so on.

Natural breaks are a graphic way of determining categories by examining the data or the histogram and looking for breaks in the frequencies or change of slope in graphs. This results in a natural grouping of similar values (Figure 8.32). This method is often used by teachers in assigning student grades. For mapping it can produce categories that are very close to the data model.

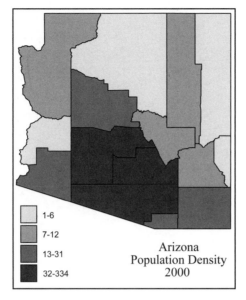

FIGURE 8.29. Choropleth map using quantiles.

Symbolizing Volumes with Lines

If data have been gathered at *sampling points*, usually called *control points*, for a continuous distribution, such as weather stations for climate data, the statistical surface will be smooth and undulating, and a form of the *isarithm* will be a suitable method of symbolizing the phenomena.

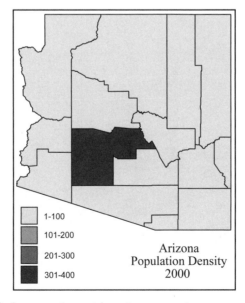

FIGURE 8.30. Choropleth map using arithmetic progression.

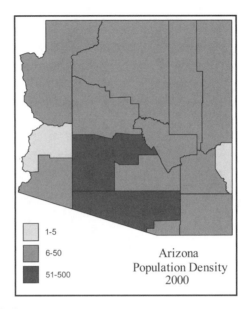

FIGURE 8.31. Choropleth map using geometric progression.

*Isarithm*s or isarithmic lines, often called *isolines*, is a generic term for a line that joins all points that have the same value above or below some datum or starting point. There are a great many kinds of isarithmic lines, most of which have specific names, although all are based on the same concept. Some of the commonly encountered lines are *isotherms* (lines of equal temperature), *isohyets* (lines of equal rainfall), *isobars* (lines of equal barometric pressure), *isobaths* (lines of equal depth below

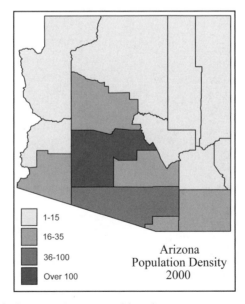

FIGURE 8.32 . Choropleth map using natural breaks.

...ve sea level, also called *contour*

...istical surface, the line of inter-
...downward is an isarithmic line
...he intersecting line is the same
...s intersects the surface, the lines
...n the planes is called the *isarith-*

...ic lines and isoplethic lines, and
...nore commonly used, it helps to
...categories. These categories are
...which the data are collected. If
...her stations, or elevation control
..., the type of line is an *isometric*
...emperature at a specific time, or
...oint.
...d are assumed or arbitrary (con-
...or the center of an enumeration
...o an area rather than a point, and
...d percentages. This type of line is
...ed to represent population densi-

...o kinds of isarithmic line are simi-
...what the reader can extract from
...lues to be determined by *interpo-*

...mparing the distance to all other
...assumes an even gradient between
...all line on Figure 8.34 is midway
...c maps, the reader interpolates or
...iny point.
...ange is found from the spacing of

No Tea For Me!
Say No to Fear and Hate
www.pfaw.org

PEOPLE FOR THE AMERICAN WAY.

TABLE 8.3. Commonly Used Isarithms

Isobars	Lines of equal barometric pressure
Isobaths	Lines of equal water depth
Isochrones	Lines of equal minimum travel time from a point
Isogones	Lines of equal earth magnetism (declination)
Isohyets	Lines of equal precipitation
Isohypses	Lines of equal land elevation (more commonly called "contour lines")
Isotherms	Lines of equal temperature

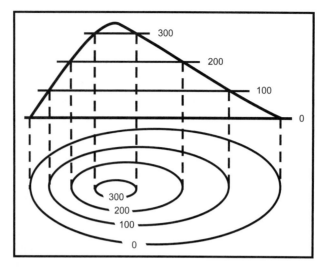

FIGURE 8.33. Formation of isarithms.

the isarithms. Evenly spaced lines represent an even rate, closely spaced lines represent a steep gradient, and widely spaced lines represent a gentle gradient.

The population map in Figure 8.35, while similar in appearance to the rainfall map, is an isoplethic map. The isopleth does not permit the estimation of population at any point because the control points represent values for areas, not actual points. That is, the values describe population per square mile, not numbers of people at specific points. This map does, however, permit the reader to note patterns and general gradients. In the same way that rate of change can be determined from an isometric map, areas of rapid or gradual change can be found on isoplethic maps.

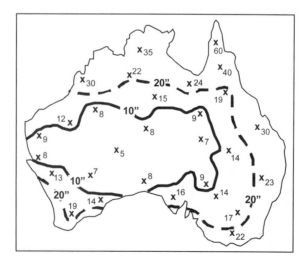

FIGURE 8.34. Drawing an isohyet (lines of equal rainfall) uses interpolation.

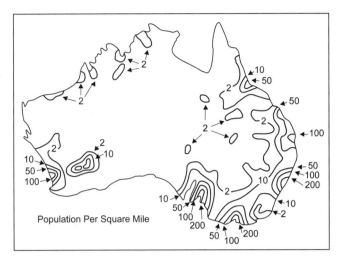

FIGURE 8.35. Isopleth map showing population density for Australia.

Creating Isarithms

Before data collection begins, a researcher selects control points for which the data are gathered. For isometric maps, these points may already exist, such as weather stations, but for isoplethic maps the points are conceptual and must be chosen. There are many options, and the choice of control point makes a difference in the look of the map. Three examples are centroid of an enumeration area, center of the data distribution, and some known place, such as a capital city or county seat. As can be seen from Figures 8.36a, 8.36b, and 8.36c, three different population maps of Arizona result from these three choices. For some kinds of studies, especially biogeographic studies, the points may be simply regularly spaced. In this case, you should avoid a rectangular pattern because it results in ambiguous situations (see Figure 8.37). Triangular or hexagonal arrangements work best.

A second decision is the isarithmic interval. For isometric maps, ideally, the interval chosen is even—for example, 10, 20, 30 or 20, 40, 60. This makes the map easier to interpret since the spacing of the lines tells the reader the nature of the surface. If an uneven interval is chosen, the line spacing can be misleading (see Figure 8.38). For isoplethic maps the data often have sharp differences and wide ranges, making it difficult to choose an even interval. In this case, the uneven interval must be clearly shown in the legend. The techniques for choosing categories for choropleth maps may be used for isopleth maps.

Design Considerations

The line widths for isarithmic maps are uniform; thus, the major design consideration is whether to shade between the lines and what scheme to use for shading. If other information will be shown on the map in addition to the isarithms, as on topographic maps, the lines are simply numbered and there is no other variable; however, it the data represented by the lines are the primary subject of the map, shading called *layer*

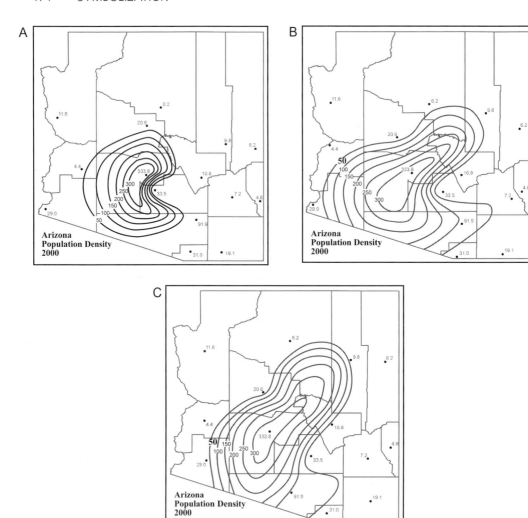

FIGURE 8.36. (A) Centers of population used as control points; (B) county seats used as control points; (C) centers of counties used as control points.

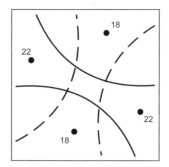

FIGURE 8.37. Rectangular arrangements of control points can result in ambiguous situations. Which are the correct 20 lines, the dotted lines or the solid lines? This can be important when designing a research study.

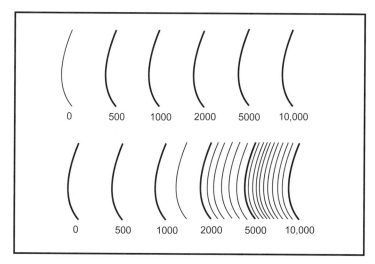

FIGURE 8.38. The top lines are drawn with an uneven isarithmic interval. Note the difference when an even interval is selected in the diagram below.

tints is often added to make the map easier to read. (If the map shows elevations, the lines are commonly called *contour lines* and the tints are called *hypsometric tints*.) The options for layer tints are lightness (tonal value), with darker shades representing greater amounts (the convention); double-ended; and spectral sequences (Figure 8.39 and Plate 4.9).

Temperature maps commonly use a double-ended sequence of reds and blues. A variation of the spectral sequence is commonly used for elevation maps with cool colors, usually greens, representing low elevations and warm colors, usually reds or reddish browns, representing high elevations. The theory behind this is that cool colors appear to recede and look farther away and warm colors appear to be closer.

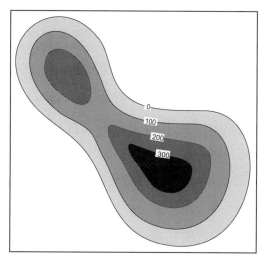

FIGURE 8.39. Layer tints are often used with isarithmic lines.

While this isn't a strong correlation, this palette has become conventional. There are variations on this sequence, however. Some commercial firms use white for the highest elevations, giving the impression of snow-capped mountains, and some have used purple, presumably based on the appearance of mountains seen in the distance. Frequently, relief shading is added to the hypsometric tints.

LEGENDS FOR ISARITHMIC MAPS

If the isarithmic interval is even, lines are identified on the map, and no layer tints are used, a simple statement of isarithmic interval can be enough, as in Figure 8.25; however, if layer tints are used, a legend showing the value scale is required. Because it is the lines that are important, not the shading, the ramp (lightness scale) is best shown by adjoining tones separated by lines and the line value marked on the line, as in Figure 8.25. It is not appropriate to show categories such as 10–20, 21–30, as on a choropleth map, because that gives an incorrect impression that the shades and categories are most important rather than the values. The tints are there simply to aid the reader in finding and interpreting the isarithms. Separate boxes should not be used in any case.

CHOOSING SYMBOL TYPES

Whether one chooses quantitative points, choropleths, isarithms, or qualitative point, line or area symbols depends on several factors that were described in Chapter 7. These factors are the nature of the phenomena, the form of the data, the appropriate visual variables, and the tools available. These factors are summarized in Table 8.4.

SUGGESTIONS FOR FURTHER READING

Dorling, Daniel, and Fairbairn, David. (1997). *Mapping: Ways of Representing the World.* New York: Prentice Hall.

Kraak, Manno-Jan, and Ormeling, Ferjan. (2003). *Cartography: Visualization of Geospatial Data* (2nd ed.). New York: Prentice Hall.

MacEachren, Alan M. (1994). *Some Truth with Maps: A Primer on Symbolization and Design.* Washington, DC: Association of American Geographers.

MacEachren, Alan M. (2004). *How Maps Work: Representation, Visualization, and Design.* New York: Guilford Press.

Tobler, Waldo. (1973). Choropleth Maps without Class Intervals. *Geographical Analysis, 5,* 262–265.

TABLE 8.4. Using Visual Variables

Type	Visual variables	Data	Nature of phenomena	Notes
Qualitative point	Shape, hue	Qualitative	Point	
Qualitative line	Shape, hue, pattern	Qualitative	Linear	
Qualitative area	Pattern, hue, orientation	Qualitative	Areal	
Dot	Location, hue, shape	Quantitative totals	Point	Additional location information required
Dot density	Location	Quantitative totals	Point	Best if enumeration areas are small
Proportional circle	Size	Interval or ratio, totals	Point, area	
Range-graded figure	Size	Totals, ordinal, interval, or ratio	Point, area	
Choropleth	Lightness, texture	Interval or ratio, derived	Volume	
Dasymetric choropleth	Lightness, texture	Interval or ratio, derived,	Volume	Additional information needed
Unclassed choropleth	Lightness, texture	Interval, ratio, derived	Volume	
Isometric	Hue, lightness pattern with texture (marginal)	True or actual points, actual or derived data	Volume	Symbol itself is a line, but shading can be added between the lines
Isopleth	Color value, pattern with texture (marginal)	Conceptual points, derived data	Volume	Shading added between the lines

Chapter 9

Multivariate Mapping

> . . . for all the interesting worlds (physical, biological, imaginary, human) that we seek to understand are inevitably and happily multivariate in nature.
>
> —EDWARD R. TUFTE, *Envisioning Information* (1990)

Often we need to show more than one variable. If the variables are unrelated, then two separate maps might be needed, but if the variables are related in some way and the goal is to show that relationship, then two or more variables can be combined on one map.

One of the simplest ways of representing two variables is to combine two conventional symbols on one map. The data from Figure 9.1a and 9.1b are combined in Figure 9.2 as choropleth and graduated circle symbols. Some symbols are specifically designed to represent multiple variables.

Some multivariate maps represent the variables with graphs or point symbols and others utilize colors in much the same way as conventional choropleths and qualitative area maps.

POINT SYMBOLS

Dot Maps

We have seen how dots on maps can show distributions of people and products. The examples given were for one variable. However, two different products or groups can be shown by varying the shape or the hue of the dots, as in Figure 9.3. The locations of the dots show the areas of density for each product or group and the areas of overlap. This technique is generally considered best for products that have comparatively little overlap so that the individual distributions can be seen; spring and winter wheat are an example. On black-and-white maps, the dots can be two shades of gray or

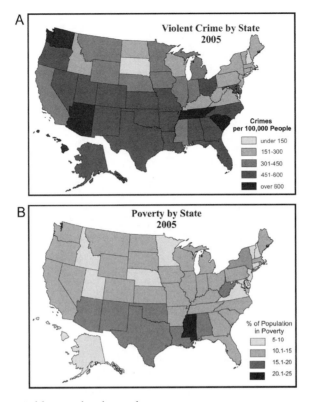

FIGURE 9.1. Two variables can be shown by two maps.

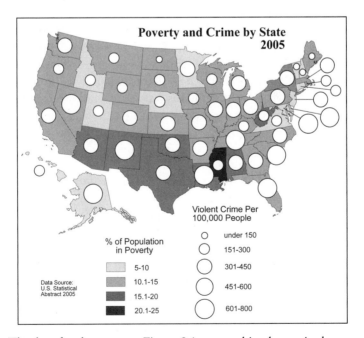

FIGURE 9.2. The data for the maps on Figure 9.1 are combined on a single map using choropleth for poverty and proportional circles for crime.

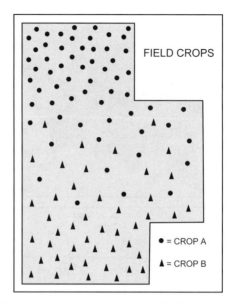

FIGURE 9.3. Different shaped dots can be used for different products.

open versus solid circles, or the shape of the dots can be varied, such as a circle and a square or a triangle. However, with small symbols it can be difficult to distinguish between solid circles and solid squares, so multiple dots work best with color maps.

Other dot methods have been suggested for showing multiple variables, such as the *pointillism technique* put forward in the 1950s by George Jenks that employs different colored dots to represent different crops or products in areas of overlap. Jenks reasoned that map readers would see the dots blending and visualize areas of transition. In the 1990s, Alan MacEachren and David DiBiase suggested the *chorodot*, which used small shaded squares within enumeration areas. The shading represents quantities, as on choropleth maps, but the squares themselves do not represent enumeration areas. Neither of these techniques has been widely used, but both might bear revisiting with increased use of animation.

Pie Charts

More properly called *segmented circles* or *segmented proportional circles*, pie charts are one of the oldest and most frequently seen multivariate symbol. In the most simple form, commonly seen as graphs (Figure 9.4), pie charts are circles with wedges representing the variables. On many presentations of nonspatial data, only one circle is used, which represents a total, and the wedges represent fractions or percentages of that total. When used on maps, the circles vary in size with a total and they are placed in the centers of the enumeration areas or at the data point just as a common proportional circle is. As with single variable circles, the size of the circle may be truly proportional or range-graded, but truly proportional symbols provide the greatest amount of information. The wedges represent fractions or percentages of the total and are used to compare categories (Figure 9.5).

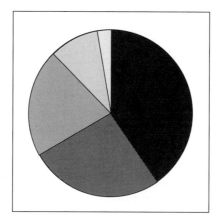

FIGURE 9.4. Pie charts use wedges to show different variables.

Design Considerations

When designing pie charts, the wedges start at the 12 o'clock position and keep their same relative positions, that is, they don't change location with changing size or importance. Of course, colors chosen for the wedges must harmonize with one another; a wedge that is consistently small might be shown in a lighter or more intense color. The 12 o'clock position is oriented to the margins of the map; if meridians are shown or

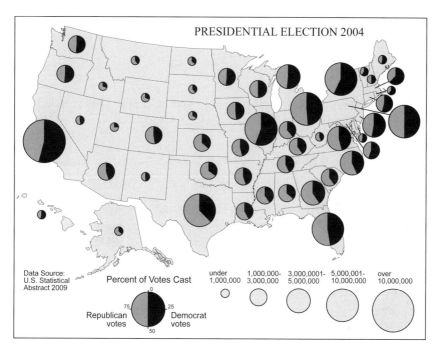

FIGURE 9.5. On a map the pie charts can vary in size to represent totals and the wedges represent different categories or products. The circles can be oriented to the meridians or to the map borders.

implied, as with some state boundaries, it can be oriented to the meridians. The legend should indicate whether the circles are range-graded or truly proportional and also whether the wedges represent percentages of the total or fractions of the total.

Chernoff Faces

Chernoff faces (Figure 9.6; Plate 9.1) were first introduced by Herman Chernoff in 1973. These are a way to show multivariate data and complex relationships in an easy-to-grasp, eye-catching manner. With this symbol, one can show as many as 15 variables by using variations of eyes, nose, mouth, shape of head, size and placement of ears, and color of face. The award-winning map by Eugene Turner (Plate 9.1) shows four variables using eyes, mouth, head shape, and color. A number of computer programs are now available, including ArcGIS, to aid in creating Chernoff faces.

Problems with Chernoff Faces

Chernoff faces are eye-catching and can be effective. However, they have two potential problems. One is that of perceived stereotyping or bias when color is used on the faces. If black or white is used to signify poverty or crime, critics complain that races are being stereotyped; the counterargument is that green or purple faces, which have been used, are jarring and perhaps give the impression of extraterrestrials. Therefore, color should be used with care. Another problem that can arise is that of too much information. If faces with multiple variables are used on a map with dozens of enumeration areas, the map becomes hard to comprehend. No studies have been done to determine the optimum number of variables, but intuitively, a smaller number of variables would seem to be more understandable than a greater number.

Other Graphs at Points

Several other less common point symbols are used on maps. Often called "icons" or "glyphs," they are all used for comparison of quantitative attributes of three or more variables.

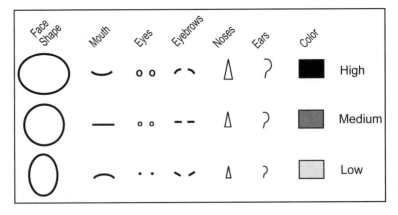

FIGURE 9.6. Chernoff faces can represent up to 15 variables. The chart here shows seven.

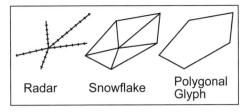

FIGURE 9.7. Graphs and glyphs.

Radar Graphs, Snowflakes, Stars, and Polygonal Glyphs

Radar graphs, also called *star graphs* or *spider graphs*, are graphs used for comparisons of quantities. Each ray represents a different category and tick marks on the rays indicate measurements of amounts (Figure 9.7). Unlike pie charts, the length of the ray, not the size of the wedge, is important. *Ray glyphs* simply show the radii, with each ray drawn proportional to the attribute without tick marks. *Snowflakes* are created by joining the end points of rays on radar graphs and eliminating the tick marks; *polygonal glyphs* join the end points of the rays to create irregular polygons, but the rays are eliminated. It is important to remember that for all of these graphs, it is the length of the rays that is important, not the area or size of the segments.

Climate Graphs

Line and bar graphs called *climographs* representing climate information of temperature and rainfall are often used on maps to show the nature of the temperature and rainfall regime for a place (Figure 9.8). These may be included as illustrations surrounding the map proper as exemplars of climate types or miniature versions may be placed at points on the map to show climate at specific places.

Any of these graphs may be combined with other symbols, such as choropleths or isarithms, to add additional variables. However, there is danger of creating a cluttered, hard-to-read map if too many variables and symbols are used. The purpose of these point graphs is to compare variables and make the information easy to grasp; too much information defeats this purpose. Multiple maps might be more user-friendly.

AREA SYMBOLS

Bivariate and Trivariate Choropleth

Choropleth maps can show two or three variables by using blends of colors or lines. These work best when there is a definite relationship between the variables, but when looking for relationships, they can be used as a part of the visualization process. Black-and-white bivariate choropleth maps employ line patterns where the spacing of the lines indicates quantities or relative amounts (Figure 9.9). It is rare to see this map type now because color is comparatively easy to use. Although the bivariate maps most commonly seen are quantitative, qualitative bivariate maps can also be constructed that show, for example, land use or crops.

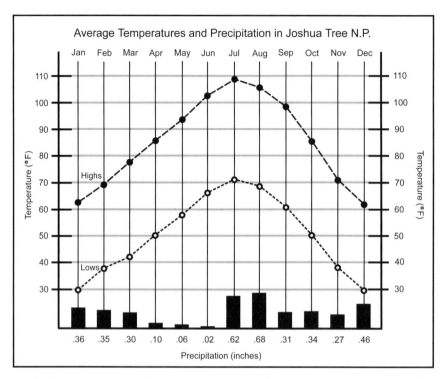

FIGURE 9.8. Climographs can be used as point symbols on a map or they can be placed outside the subject area.

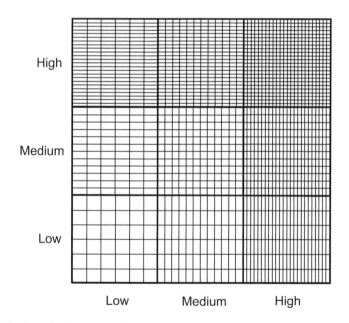

FIGURE 9.9. Black-and-white bivariate choropleth chart.

The color choices are described more fully in Chapter 4, but the basic concept is shown in Figure 9.9 and Plate 9.2. In general, the color schemes used are pairs of sequential schemes, pairs of diverging schemes, or paired sequential and diverging schemes. In Plate 9.2 the simplest form is shown, a paired sequential scheme, using only two hues that vary in lightness. The three subtractive primaries of cyan, yellow, and magenta are used to create nine different combinations. Figure 9.10 shows the percentage of each primary used in order to obtain the nine colors. This scheme is easy to understand and create. More sophisticated schemes can be made by varying saturation as well as lightness.

Most commonly, the legend on bivariate choropleth maps takes the form of a rectangular grid on which positive and negative diagonals progress from low to high or high to low.

Judy Olson has provided guidelines for creating bivariate choropleths that were summarized by J. Ronald Eyton (1984):

1. The colors should be distinguishable and the transitions should be smooth and visually logical.

2. One should be able to distinguish individual categories, and one should be able to differentiate the two distributions.

3. The arrangement of colors in the legend should correspond to the arrangement of a scatter diagram of the distribution.

4. The colors should progress in lightness from light to dark with high data values represented by dark tones and low data values by light tones.

5. To convey a relationship, the positive and negative diagonals should have visual coherence.

6. The bivariate map should be constructed as a direct combination of the specific sets of colors assigned to the two individual maps.

7. The combination of colors on the two individual maps should look like combinations of the specific colors involved.

8. The number of categories should not be so large that the reader is overwhelmed.

100 C 0 M 100 Y	100 C 50 M 50 Y	100 C 100 M 0 Y
50 C 0 M 100 Y	50 C 50 M 50 Y	50 C 100 M 0 Y
0 C 0 M 100 Y	0 C 50 M 50 Y	0 C 100 M 0 Y

FIGURE 9.10. Simple color scheme for a bivariate choropleth. See also Plates 9.2 and 9.3.

Because of the complexity of the map and legend, a three-by-three legend is easier to comprehend than a four-by-four or five-by-five. A five-by-five legend requires the reader to compare 25 different colors and understand their meanings.

Plate 9.3 is a simple bivariate map that combines the information on Figures 9.1a and 9.1b.

While three-variable (*trivariate*) choropleth maps can be made, they are less common because they are more difficult for the reader to understand. Plate 9.4 is an example of a trivariate color chart for a map on which the hues are the three subtractive primaries.

Because multivariate maps are more complex than single-variable maps, the decision to create such maps should take into account the probable map-reading sophistication of the reader and whether the purpose of the map is to provide detailed information or general patterns. Specifically, will the patterns and relationships presented be shown effectively with a multivariate map, or will a map series be more useful?

SUGGESTIONS FOR FURTHER READING

Brewer, Cynthia A. (1994). Color Use Guidelines for Mapping and Visualization. In Alan M. MacEachren and D. R. Fraser Taylor (Eds.), *Visualization in Modern Cartography* (pp. 123–147). New York: Pergamon.

Eyton, J. Ronald. (1984). Map Supplement: Complementary-Color, Two-Variable Maps. *Annals of the Association of American Geographers, 74*, 477–490.

Harris, Robert L. (1999). *Information Graphics: A Comprehensive Illustrated Reference.* New York: Oxford University Press.

Olson, Judy M. (1981). Spectrally Encoded Two-Variable Maps. *Annals of the Association of American Geographers, 71*, 259–276.

Slocum, Terry A., et al. (2005). *Thematic Cartography and Geographic Visualization* (2nd ed.). Upper Saddle River, NJ: Pearson Prentice Hall.

Tufte, Edward R. (1983). *The Visual Display of Quantitative Information.* Cheshire, CT: Graphics Press.

Tufte, Edward R. (1990). *Envisioning Information.* Cheshire, CT: Graphics Press.

PART IV

NONTRADITIONAL MAPPING

Chapter 10

Cartograms and Diagrams

Our socioeconomic overview of the world will be more realistic if
we think of the relative importance of its parts in the proportions
of a population cartogram rather than in the proportions of a map.
The results are often quite surprising. Such cartograms can make
certain problems startlingly clear. . . .

—ERWIN RAISZ, *Principles of
Cartography* (1962)

CARTOGRAMS

Cartograms of various kinds are becoming increasingly popular, in part perhaps
because computer techniques make them easier to construct. A *cartogram* is a geographic representation on which size or distance is scaled to a variable other than
earth size or distance units. Thus, the map may be scaled to population, or time, or
costs.

Cartograms have existed since the 19th century when statistical maps became
popular. One of the earliest, which was published in a school atlas, was unusual in
that it showed countries as rectangles representing actual areas of countries rather
than a variable such as population and was designed to let children compare areas. It
has been called an *area cartogram*, but might more properly be called a *diagram* or
geoschematic. Erwin Raisz, in 1934, created a series of cartograms and helped popu-
larize the form. Figure 10.1, drawn by Raisz, is a rectangular cartogram on which the
size of states are shown according to population; this is a true cartogram.

Cartogram Types

Several kinds of cartograms are in common use. The most commonly seen are variations of the *value-by-area cartogram* in which the size of enumeration areas vary
according to the value represented, such as population, income, electoral votes, or
median age, not with the actual geographic area. (The Raisz example is a value-

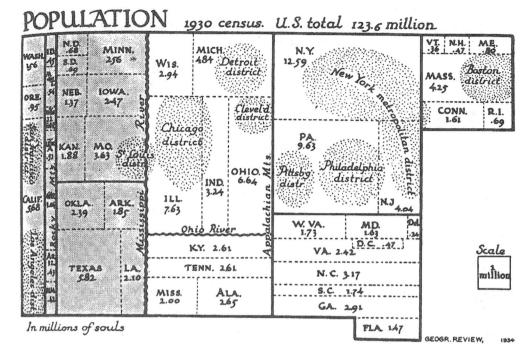

FIGURE 10.1. Erwin Raisz rectangular cartogram. From Raisz, Erwin. (1934). "The Rectangular Statistical Cartograms. *Geographical Review*, 24, 292–296. Reprinted with permission from the American Geographical Society.

by-area cartogram.) Within this type are several subcategories that are discussed below.

A second type of cartogram uses a time scale instead of a distance scale; these are sometimes called *distance cartograms* or *linear cartograms*. At one time, any spatial diagram or model was called a cartogram; Raisz described "timetable cartograms" that showed transportation routes as a cartogram, but they are more properly called diagrams.

Value-by-Area Cartograms

CONTIGUOUS CARTOGRAMS

Contiguous cartograms attempt to maintain borders between enumeration areas although shapes are distorted. The *rectangular cartogram* maintains borders, but not shapes; the shapes of areas become squares and rectangles.

A type of contiguous cartogram that has become more popular since the advent of computers is the topologically correct cartogram. This was a very difficult task to perform before the advent of computers. Figure 10.2 is a hand-drawn cartogram, and Plate 10.1 is a computer-generated cartogram. The maps are highly abstracted representations on which actual locations or outlines are distorted and even the relationships between areas may be distorted. Although the border relationships are maintained on Plate 10.1, the shapes are badly distorted and in some cases unrecognizable.

Advantages and Disadvantages. Research, especially by the late Borden Dent, has shown that a contiguous cartogram that approximates shape with straight line segments is probably the most useful and the least confusing to the reader. His studies showed that of the qualities exhibited by a conventional geographic base map— shape, orientation, and contiguity—shape is the essential factor to preserve on a cartogram.

Shape aids the reader in retrieving information from the map; if individual states cannot be recognized and compared with either a conventional map or a mental map of the geographic space, then the cartogram can have no effect. Dent felt a value-by-area cartogram should not be used if the general shape of the enumeration areas cannot be maintained. It might also be noted that if the shapes of the enumeration areas are unfamiliar to the reader, the cartogram loses its effectiveness. Labeling or including an inset with a conventional map might aid the reader, but more standard representations, such as census tracts or minor civil divisions, should be considered for unfamiliar areas.

Cartograms that are drawn with curved outlines and are truly contiguous must distort shapes excessively and are therefore more difficult for the reader to interpret unless only a few very familiar areas are used. True contiguity appears to be less

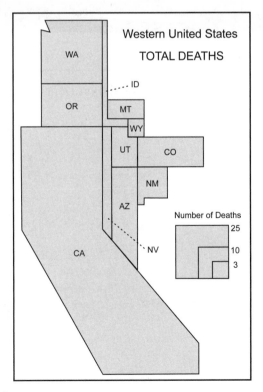

FIGURE 10.2. Hand-drawn cartogram.

important than shape for reader acceptance and recognition. However, probably the most commonly seen cartograms, especially on the Internet, do not preserve shape.

NONCONTIGUOUS CARTOGRAMS

Traditional noncontiguous value-by-area cartograms show shapes correctly and either enlarge or reduce enumeration areas according to the variable represented; adjoining areas do not touch (Figure 10.3). Although there is usually an effort to place the various areas in roughly their correct positions relative to one another, they are separated by empty spaces. Often, as in Figure 10.3, a true outline of the area is placed around the enumeration areas. This gives the reader a frame of reference. It allows the reader to see the region in its entirety in true geographical space and gives an idea of the gaps, which are considered important in this type of cartogram.

Advantages and Disadvantages. Judy Olson (1976) lists three properties of the noncontiguous cartogram that make it a useful device:

1. [The gaps between the enumeration units are] meaningful representations of discrepancies of values.

2. The representation and manipulations involve only the discrete units for which information is available and only the lines that can be accurately relocated on the original map appear on the noncontiguous cartogram.

3. Because only sizes of units change, not their shapes, recognition of the units represented is relatively uncomplicated for the reader.

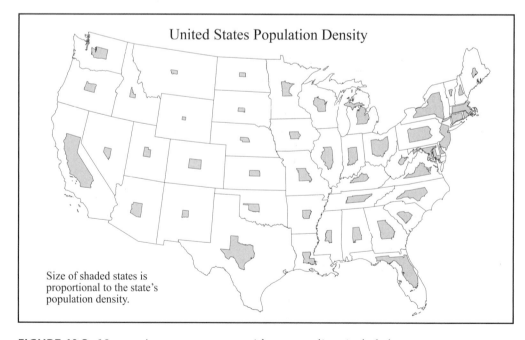

United States Population Density

Size of shaded states is proportional to the state's population density.

FIGURE 10.3. Noncontiguous cartogram with state outlines included.

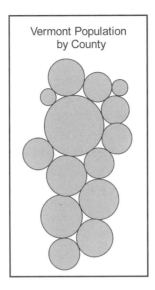

Vermont Population
by County

FIGURE 10.4. Dorling cartogram.

Some authors consider the noncontiguous type to be the most crude form of the cartogram and criticize the sacrifice of continuity. However, it is easy to create even with illustration software.

VARIATIONS

There are a number of variations of value-by-area cartograms. The rectangular cartogram, popularized by Raisz, uses a variety of rectangles to approximate the shapes of countries or states. This type is popular in atlases. *Dorling cartograms,* which were invented by Daniel Dorling, replace the enumeration areas by uniform abstract shapes, normally circles (Figure 10.4); there is no attempt to maintain shape or topology. It is best if the shapes do not overlap so that each enumeration area can be identified, but in doing so, the figures are moved from their actual geographic locations. The method for determining the circles or squares is the same as that used for proportional circles and squares. The *Demers cartogram* is a variation of the Dorling cartogram, but uses squares to represent enumeration areas and permits greater contiguity than circles (Figure 10.5).

Bivariate Cartograms

Any type of value-by-area cartogram can represent multiple variables. *Bivariate cartograms*, like bivariate choropleth maps, show two variables, such as population and income. One variable is shown as the cartogram, while the second variable is shown by color or shading, as with choropleth maps Figure 10.6 is a simple, noncontiguous cartogram in which the population of each state forms the cartogram, and violent crime is shown with shades of gray.

Finally, contiguous cartograms likewise portray the enumeration area as the car-

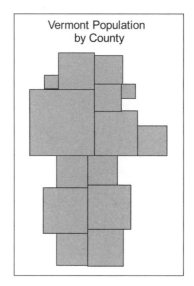

FIGURE 10.5. Demers cartogram.

togram and add color or shading to represent the second variable. This can be done with rectangular cartograms, as in many atlases, or with topologically correct cartograms, which are popular on the Internet. These were especially popular during the U.S. national elections of 2008 (Plate 10.1). While these are dramatic and eye-catching, more studies on their effectiveness need to be carried out.

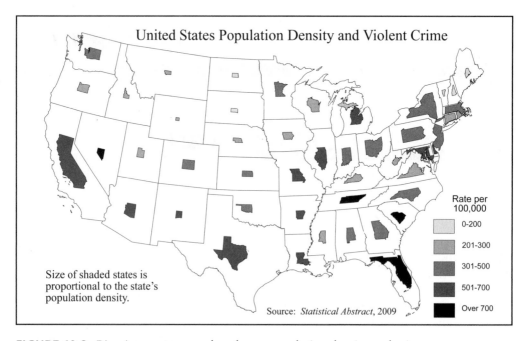

FIGURE 10.6. Bivariate cartogram that shows population density and crime rate.

It is also possible to create multivariate cartograms that combine a cartogram with the principles of the bivariate choropleth.

Distance-by-Time Cartograms

Distance-by-time cartograms have also been called *linear cartograms* and *distance cartograms*, but since conventional maps are scaled according to distance units, the latter is somewhat inaccurate and misleading. These cartograms vary map distances according to the time needed to travel the real-world distance. This is not a new concept; some primitive peoples normally made their maps according to a time scale, rather than a distance scale, because the length of time needed to travel between two points was of more importance than the actual geographic distance. Today, anyone who has traveled an expressway or freeway route during light traffic and rush-hour traffic can appreciate the value of expressing distances in terms of time. Two places can be exactly the same number of miles from a given point, but one might be 1 hour away and the other 30 minutes away because of the nature of the roads, amount of traffic, presence or absence of traffic lights, and nature of the terrain. In fact, on some freeways the time to get to the next interchange or the airport is posted on flashing roadside signs.

Two kinds of cartograms are used to represent time; one is linear and shows time from point to point and may be either directional or not, and the other shows time from a center point.

Figure 10.7 illustrates a linear cartogram scaled for time along a specific route. Walking time is used as the scale. It can be seen that uphill segments are longer than downhill segments. These cartograms are most useful if they show direction of travel;

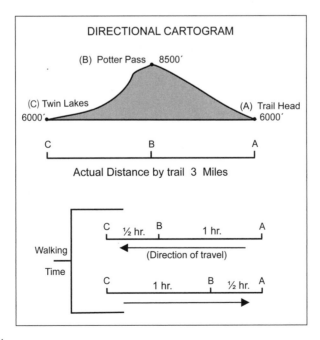

FIGURE 10.7. Linear cartogram.

otherwise, time must be averaged in both directions, and this is inaccurate. The figure shows the route in two directional segments; segment A–B going westward is longer than the same segment going eastward because it is uphill on the westward journey and downhill on the eastward journey. Such charts can be a useful addition to hiking trail maps.

These cartograms are simple to construct; the most time-consuming aspect is the fieldwork because seldom does one find charts showing travel time. The route must be traversed, preferably more than once, to determine the average travel time. A scale is chosen, for example, 1 inch represents 1 hour, and usually the route itself is highly generalized.

The *center point cartogram* (Figure 10.8) shows times from a center point and the cartogram can be read only that way. Direction is important; times to and from the center point might differ significantly. Times cannot be determined between any other points on the map; they are all related to the center. The time scale can be for average times or only apply to a specific time of day if freeway routes are the subject. A center point cartogram shows the same information as an isochronic (lines of equal time) map, but the isochrones are converted to concentric circles.

The first step in creating a center point cartogram is construction of an isochronic map. Travel times must be determined from the center point to all relevant points. An isochronic line connects all points the same time away from the center (Figure 10.9). These lines are then converted to concentric circles that distort the geographic base by removing some locations from the center and pulling others closer. On some of these maps the distorted base is shown; on others only the concentric circles are shown with place-names labeled to act as orienting points for the reader. Another variation of the distance-by-time cartogram simply uses lines radiating from the center point to the other points of interest. Again, the geographic base is distorted.

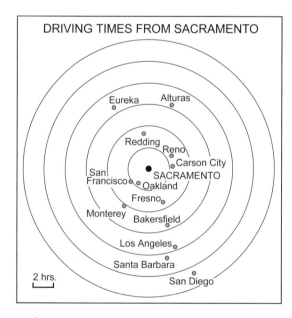

FIGURE 10.8. Center point cartogram.

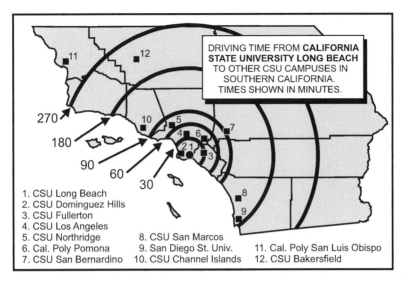

FIGURE 10.9. Variation on an isochronic (equal time) map. The lines represent driving time and the locations are in their correct geographic positions.

Why Use Cartograms?

Cartograms of any type have a strong visual impact. Because they represent an unusual view of the world or an area, they attract reader attention. This is an important reason for choosing a cartogram instead of a conventional map. Often we are concerned with magnitudes and want to make a stronger impression of relative values than could be made with a conventional choroplethic or isarithmic map. On a conventional population density choropleth map by state, some small states with high population density are barely noticeable, even when shaded black or red, but on a population value-by-area cartogram of the United States, some large geographic areas in the western states appear very small because of the low population densities and New Jersey becomes very large (Figure 10.3). Conventional maps must show sizes with complex symbols; therefore a graduated-circle map, for example, does not have the same impact as a cartogram.

Both kinds of value-by-area cartograms, contiguous and noncontiguous, share some characteristics or features that make them useful cartographic devices. They have a strong visual impact; they permit the representation of distributions that might be obscured by variation in enumeration sizes if mapped by conventional means; and because there is little unnecessary detail, they may offer clearer representations of some distributions.

There is no generalization of data on cartograms. There is no loss of detail through generalization into categories as there is with choroplethic maps. Although some readers and even cartographers are disturbed by the apparent lack of accuracy, cartograms are a legitimate and useful means of displaying geographic data. Of course, because of their strong visual impact, cartograms are useful when the intent of the map is to persuade.

DIAGRAMS

Models, such as those that show idealized climates (Figure 10.10) and the like, have also been described as cartograms, but are more appropriately designated as diagrams.

Many other graphic representations are created that are not as scientific as the value-by-area and distance-by-time cartograms since the distortions are not controlled or as systematic as they are in the previous examples, but they serve many of the same functions. They are attention-getting and they communicate information in a clear, uncluttered manner. Many of these are route maps or diagrams, such as one might find for subway maps (Plate 10.2). The London Underground Map created by Harry Beck in 1931 is the most famous of these diagrams and has been imitated throughout the world.. There is no effort to preserve scale, and there is only slight conformance to orientation. The main emphasis is to show the basic route and stops of several different bus, subway, or train routes. These are usually depicted in bright colors, with each color representing a different route. The hope is that even a user unfamiliar with the language can use such maps to choose the correct route and arrive at a destination without difficulty.

A variation of this type of diagram is the use of flow lines combined with very stylized routes and a distorted base.

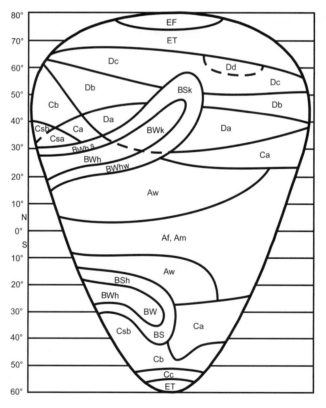

FIGURE 10.10. A model showing ideal climates.

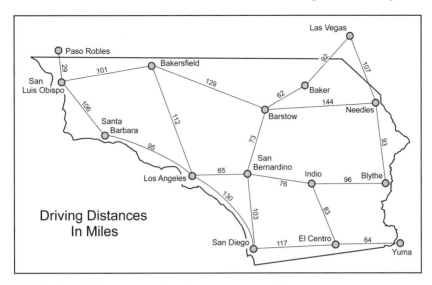

FIGURE 10.11. Road distances can be shown schematically.

Some road maps include a diagram showing distances and estimated driving times between various points, but these are not scaled according to time as are distance-by-time cartograms, and they preserve the spatial relationships between the points (Figure 10.11).

Because of the nature of these diagrams, there are no guidelines for construction; each presents a different problem. The goals are clarity, simplicity, and speed of interpretation. It is also desirable for the diagram to be converted easily into a mental map for the reader.

SUGGESTIONS FOR FURTHER READING

Cuff, David J., et al. (1984). Nested Value-by-Area Cartograms for Symbolizing Land Use and Other Proportions. *Cartographica, 21,* 1–8.

Dent, Borden. (1972). A Note on the Importance of Shape in Cartogram Communication. *Journal of Geography, 71,* 393–401.

Dent, Borden. (1975). Communication Aspect of Value-by-Area Cartograms. *American Cartographer, 2,* 154–168.

Dent, Borden, et al. (2009). *Cartography: Thematic Map Design* (6th ed.). Dubuque, IA: William C. Brown.

Olson, Judy M. (1976). Noncontiguous Area Cartograms. *Professional Geographer, 28,* 371–380.

Ovenden, Mark. (2007). *Transit Maps of the World.* London: Penguin Books.

Raisz, Erwin. (1934). Rectangular Statistical Cartogram. *Geographical Review, 24,* 292–296.

Tufte, Edward R. (1983). *The Visual Display of Quantitative Information.* Cheshire, CT: Graphics Press.

Tufte, Edward R. (1990). *Envisioning Information.* Cheshire, CT: Graphics Press.

Chapter 11

Continuity and Change in the Computer Era

> The arrival of the personal computer in the early 1980s changed production cartography forever.
>
> —MARK HARROWER, *Cartographica*
> (2004)

Although the principles described in the previous chapters apply to all kinds of maps regardless of the methods used to prepare them, some map types have come of age in the computer era and have special requirements that traditional paper maps do not.

Much of the change in mapping in recent years has been a result of the increasing availability and power of the personal computer. The importance of the personal computer to cartography in general has been shown throughout the book, but a number of specific map types have been developed or become popular only with computer assistance. Among these are animated maps, haptic maps, sound maps, multimedia maps, and Web maps. Because this book is focused on the design process, I address the design aspects of these maps that differ from traditional maps, but not the software used to make them. Rapid changes in software and hardware mean that anyone working in this area must keep abreast of innovations described in professional journals and on manufacturer's websites; the lag time between writing a book and its publication means that technological information in books is 1–3 years old.

Although these map types are all computer-created, some were possible before the computer age. They tend to be software-driven, that is, they require specific software, not general GIS or illustration programs; they may require specialized hardware; and they may require a team of experts rather than one person to create them. While cartographers throughout history have worked in concert with geographers and printers, now they find that computer programmers and hardware experts are necessary team members. The geographer provides content (sometimes the cartographer and the geographer are the same person), then the cartographer focuses on the design process and communicates with the computer experts on what programming is needed.

A factor in rapid change is methods of distribution. Whereas conventional maps have been distributed primarily through print media, the new generation of maps is disseminated on CD, DVD, or through the Internet, and these new maps are viewed on computer monitors. This raises a number of design and layout issues involving format and color, among others.

A problem for the map designer is that many of these map types are so new that little research has been carried out on the effectiveness of their design. Thus, there are fewer guidelines for the new cartography than for conventional cartography. Much is done simply by common practice, but that practice may only be a few years old. There are many unanswered questions. Do people learn better from animated maps? Do multimedia presentations incorporating pictures, video, sound, and animation assist the user in finding information, or do they result in sensory overload? Is there an optimal number of sensory elements that can be included in a presentation?

Additional design elements need to be worked into the composition of computer maps and some special considerations come up, such as additional graphic variables, depending on the type of map, but the same design questions asked before beginning any map apply here. In the following sections I focus on these elements and questions. Material in the Bibliography and Appendices will direct the reader to more extensive discussions of specific map types.

DIGITAL AND ONLINE MAPS

Two kinds of maps that bear mentioning are digital maps and mapping programs that are available on DVD, and online maps that are accessed through the Web. Usually there is little design involved for the user, and other than color and category choices, there is little or no design flexibility.

MapPoint by Microsoft is a basic GIS program on DVD geared toward business that allows data analysis from user data as well as some built-in census data. Choropleth, graduated circles, shaded graduated circles, pie charts, both uniform and sized, and "push pins" that allow one-to-one point mapping are possible. A set of simple drawing tools is provided so that the user can circle areas and draw arrows and the like, but the basic map layout is fixed. It is possible to copy the map to an illustration program and work with the design, but probably most users are more interested in data analysis than presentation.

Online maps are found on the Internet and include such sites as MapQuest, Google Maps, Google Earth, and Navteq. Although some, notably Google Earth, have some mapping capabilities, most often these maps are used for location and route finding. A list of these sites is found in Appendix B, but because little designing is done by most users, the maps are not covered in detail here. Numerous how-to books are available for producing maps based on Google Earth.

ANIMATED MAPS

Animated cartography has a long history, with research papers on the subject dating to the 1950s; simple animated maps that showed such things as moving arrows on maps were made before and during World War II. There is an extensive literature on these maps; Michael P. Peterson wrote the first book on animated cartography in 1995, although Norman Thrower urged cartographers to explore the possibilities of animation as early as 1959, and in the 1970s Waldo Tobler and Hal Moellering produced early map animations by computer. While animated maps predate the computer era, they were little used because they were time-consuming to produce and could only be viewed on film, effectively limiting them to motion pictures, especially newsreels, and educational documentaries to be viewed in schools. With the advent of powerful, high-performance personal computers, commercial animation software, CDs and DVDs, and the Internet, such maps became easier to create, view, and disseminate.

Animations basically show change and especially change through space and time, therefore adding a fourth dimension to maps. There are two basic kinds of animations: *temporal*, which show change through time, and *nontemporal*, which are not time-related, but may show change in space or some attribute. Time change shows such things as spread of disease, shrinking rain forests, or changing temperatures or rainfall over a period of time. Spatial change is exemplified by the so-called flybys or flythrough animations that give the feel of flying over an area. Attribute change often focuses on points and can show such things as earthquake magnitudes or numbers of deaths in a war with flashing points or choropleth maps showing different age groups.

Animated cartography may be used in both visualization and communication. The viewer who is able to fly around a 3-D data representation can see information not seen from one or even two perspectives. Looking at an animated sequence of data can reveal patterns not seen on static maps. Examining data and searching for patterns and unknowns is at the heart of the visualization process. Animations can also be used to explain knowns, tell a story, or present information; used in this way animation is a communication tool.

One must ask the same questions before beginning an animated map as for a static map, that is, what is the purpose of the map; who is the audience and what are their needs and abilities; what are the restrictions of format; and how will the map be produced and distributed?

The most important question to ask before creating an animated map is whether an animation is the best choice or merely an eye-catching gimmick? There should be some purpose to an animation, not merely the "gee whiz factor"; many animated maps would have been far more effective as a series of side-by-side maps (sequenced maps) that can be compared leisurely, rather than as three or four maps that flash before the viewers' eyes. If change is not an aspect of the data examination or presentation, animation may not be the best choice. Animated maps are still time-consuming to produce and therefore expensive, so one must take the cost factor into account. Mark Harrower (2009) has described this as the "effort-to-reward ratio."

A second question is, What is the purpose of the map? Will it be made for one person or a team examining a data set to look for spatiotemporal patterns, that is, visualization, or will it be used to communicate information?

When creating animated maps there are additional graphic variables (see Chapter 7) to consider: duration, rate of change, order, display date, frequency, and synchronization (Table 11.1). *Duration* is the period of time a frame or scene remains on the screen. A frame is a single image; a scene is a group of frames. The shorter the duration, generally, the smoother the animation. *Rate of change* is the relationship between the magnitude of the change of attributes and the duration of the frame or scene. Less magnitude of change between frames or a longer duration for each frame makes a smoother animation. *Order* is the arrangement or sequence in which the frames are presented. Most commonly the order is chronological, but other arrangements can also be used, such as quantities, that may help viewers to visualize patterns. *Display date* is the time some change is started. *Frequency* is the number of recognizable stages for each unit of time and *synchronization* is the correspondence time series.

In designing the animated map page layout there are three major considerations. The map image itself, of course, is primary and should, like conventional maps, take up most of the space available. Because of limitations of screen size and performance and the need for expanded legends and interface, these maps may be smaller than desired. This has implications for the amount of information shown. Animated maps, like static maps, require a title and legend or legends. There may be a conventional legend to explain symbols and a temporal legend to track time, but a legend is also needed to navigate the map. Three types of temporal legend are shown in Figure 11.1. The final element is the interface that allows the user to interact with the animation. This provides tools to pause, fast-forward and fast-reverse, go to a specific frame, or even reorder the sequence of frames. There may also be a link to the database.

Mark Harrower (2003) provided a series of "tips" for designing an attractive, effective animated map. He noted four challenges that animated maps present. One of these is that the reader can miss important information because it flashes on the screen too quickly. This is referred to as *disappearance*. Because of this problem, many map-reading tasks that are simple with static maps, such as estimating sizes, matching colors to a legend, comparing symbols, or reading labels, are difficult for users of animated maps. Some animations include large blocks of text that show too briefly for the user to read this text and look at the map. Some of Harrower's sugges-

TABLE 11.1. Graphic Variables

Visual	Sound	Animation[a]	Tactile	Haptic
Size	Location	Duration	Volume	Pressure
Shape	Loudness	Rate of change	Size	Spatial acuity
Hue	Pitch	Order	Value	Position
Lightness	Register	Display date	Texture	Texture
Saturation	Timbre	Frequency	Form	Hardness
Pattern	Duration	Synchronization	Orientation	Temperature
Texture	Rate of change		Elevation	
Location	Order			
Orientation	Attack/decay			

[a]In addition to the visual variables.

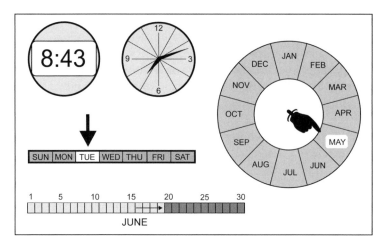

FIGURE 11.1. Temporal legends for animated maps.

tions for the problem of disappearance are showing the animation numerous times, stopping the animation and proceeding frame by frame, and adjusting the speed of the animation.

A second challenge cited by Harrower is *attention*. The user doesn't know where to look. Voice narration and sound prompts (beeps, clicks, etc.) can help direct the reader's attention. Flashing or moving symbols, such as flashing arrows pointing to a location, can help, but should be used with care, because they can become annoying.

The third challenge is *complexity*. "Many animated maps try to do too much and end up saying very little" (Harrower, 2003, p. 64). Effective animated maps are often quite generalized and focus on the most important trends or features. Instead of complex numerical legends, categories that show high, medium, and low are easier to understand. The colors used should also be distinct and easy to remember. Harrower believes that animated maps are better suited for showing geographic patterns and changes in patterns than specific rates of change.

Harrower's final challenge is *confidence*. There is some evidence that users are less confident that they understand the information they get from animated maps than from static maps. This problem can be lessened by providing an introduction to the interface before showing data. The user also needs a relatively simple interface that he or she adapts to quickly.

Design Considerations

Because of the resolution of the computer and constantly changing images, larger text is necessary for animated maps than for conventional maps. Thicker line weights, brighter colors, and more generalized base maps are needed on animated maps than on paper maps. One should also be aware that flashing lights on the map can cause problems for map users with certain medical problems.

Animated maps run the gamut from simple base maps with moving symbols to full multimedia presentations with sound and interactivity, but complexity does not necessarily mean that a map is more effective. As with all map forms, looking at many examples helps the cartographer develop an "eye" for good maps.

Unfortunately, a book is limited to static maps; animated maps cannot be shown. A few websites are listed in Appendix B. Entering "animated maps" or "cartographic animation" in a web browser will result in hundreds of examples that can be viewed and evaluated.

SOUND MAPS

Sound is usually combined with sight, touch, or both in specialized maps. Sound has the potential for being a valuable addition to maps for the visually impaired, but it can also be useful on maps for sighted readers. Sound can be used in a variety of ways on electronic maps. It can be used as (1) narrative—to identify features and describe them, common for animated maps and automobile GPS units; (2) as a memetic (realistic) symbol on a map, such as a car horn, bomb blast, or dog barking, analogous to pictorial symbols in vision; (3) as a redundant variable, such as a spoken name plus a written label; (4) instead of visual patterns; (5) as an alarm; (6) for adding nonvisual data dimensions; and (7) to reduce visual distraction or clutter by replacing labels, or by representing location.

Like visual symbols, sound can be realistic or abstract. Voice narration is the most obvious and probably most common sound used on maps. Other sounds represent real-world sounds, such as horns, doors closing, and the like. Abstract sounds can be cues or can represent data. John Krygier (1994) has proposed a basic set of elements of sound and has created a table of "sound variables" analogous to visual variables (Table 11.1). These variables are location, loudness, pitch, register, timbre, duration, rate of change, order, and attack/decay. Just as visual variables are used for qualitative (nominal) and quantitative (ordinal, interval) data, sound variables can represent data.

HAPTIC MAPS

Touch has been used for many years for maps for the visually impaired in the form of *tactual* or *tactile maps*. Such maps utilize raised symbols imprinted on a special paper or plastic that can be felt and read in a manner similar to braille writing (Figures 11.2, 11.3). *Haptic maps*, which might be described as electronic touch maps, are a new variation. A computer screen cannot be felt in the way a conventional tactual map can. With increasing use of electronic maps, those who are visually impaired are deprived of map access. Haptic maps are a way to deliver geospatial information to the blind and partially sighted with the computer. The sensations are delivered through the mouse, a wand, or a special glove. Haptics can also be used on maps for those with normal vision.

Touch may simply involve the skin, which is the type used on conventional tactile maps, or it can involve muscles, tendons, and joints. Haptics involve both of these modes and it is associated with the "feel" of objects. We can consider some of the sensations of touch to be another kind of variable, akin to visual variables; these

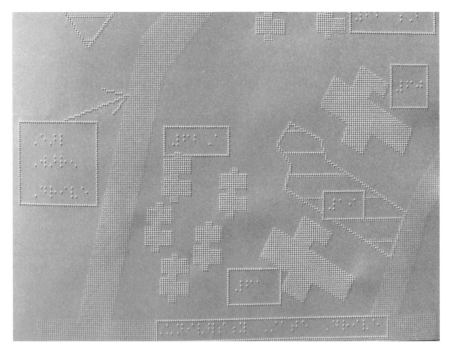

FIGURE 11.2. Part of a braille map of a college campus. Courtesy of disabled students, California State University, Long Beach.

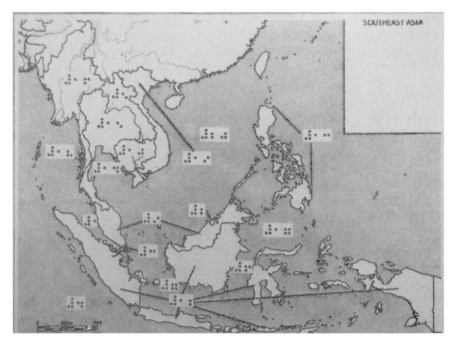

FIGURE 11.3. Braille map of Southeast Asia. Courtesy of disabled students, California State University, Long Beach.

include pressure, spatial acuity, and position. Others that provide sensory information are heat and cold, and pain, but these have not yet been used in haptic maps. It is unlikely that pain would be used as a mapping variable, but certainly, if hardware is developed to support it, temperature could be a valuable addition, especially for weather and climate maps (Table 11.1).

In making maps for the visually impaired, the mapmaker must pay particular attention to audience needs and abilities. Not all visually impaired users have the same degree of blindness, and there are differences based on whether the user is blind from birth or became blind through accident or disease at a later age (adventitiously). Adventitiously blind users have memory of sight and this can be an important factor. Additionally, those who are blind through diabetic retinopathy, an eye complication of diabetes, often have a reduced sense of touch in their fingers, making braille reading difficult or impossible. Haptics and sound are an option in this case.

MULTIMEDIA MAPS

Multimedia maps might also be called *multisensory* because these maps draw on more than one sense. Thus, video, animation, sound, touch, and static maps might all be included on the "page." In a simplified way, multimedia has been used on static maps for centuries by including illustrations, text, and graphs on the same page as the map. The computer has allowed expansion to include videos, sound, and animations, and it permits interactivity with the map so that the user can focus on one or more of the elements and turn others on or off. When we speak of multimedia maps now, the assumption is that they will be viewed on a monitor. The delivery medium may be CD, DVD, or the Internet. Many maps designed for the Internet are multimedia.

The jury is still out on the effectiveness of multimedia maps. Although they may allow the user to be an active participant, and draw on many senses, there is risk of sensory overload for the viewer. Thus, when creating such maps, one must ask if they truly add to the communication or if the user is being overwhelmed with information. In fact, this question should be asked before adding any element to the presentation.

WEB MAPS

The Web is basically a means of distribution of information, which may be in the form of text, illustrations, video, and maps. Maps for the Web have become ubiquitous and are used for a variety of reasons. Local businesses have links to sites such as MapQuest or Google Maps to aid customers in finding their locations. National companies and organizations have store locators linked to maps to help customers find directions to the nearest stores. Increasingly, museums and libraries with historic map collections put maps online so users can examine them from thousands of miles away. Individuals and organizations have statistical maps and informational maps on their websites.

All of the techniques discussed earlier in this chapter apply to Web maps—especially animation, sound, and interactivity. In general there are two main categories of Web maps: static and dynamic; varying degrees of interaction are available for each.

Static maps may be existing maps that have been scanned for display on the Web. The site for the David Rumsey map collection, *www.davidrumsey.com*, is an example of this type of Web mapping (Figure 11.4). The maps are historic maps and the only "interaction" possible is panning and zooming. The maps are not "clickable," that is, one cannot click the mouse on a map and get a link to more information. Metadata for the maps is available through the database. The colors of the maps are their original colors, which may or may not translate well to Web viewing.

Other static maps are designed specifically for Web viewing that have panning and zooming as the only interactive options. The design concerns for these maps include the usual elements of title, legend, scale, and the like of traditional maps, plus the locations of pan and zoom bars and choice of color and background.

Interactive static maps are "clickable"; rolling the mouse over an area and/or clicking the mouse button provides additional information by bringing up a different layer of information. For example, rolling the mouse over buildings shown on a campus map or city map might show the name of the building or a voice might speak the name. Clicking on different areas of a choropleth map might give the data value for the enumeration area.

Dynamic Web maps also include animation and multimedia and are generally interactive. Dynamic maps are all computer-created, but some are specifically designed for the Web, whereas others may have been designed originally to be disseminated on CD or DVD.

FIGURE 11.4. The David Rumsey website allows the viewer to see antique maps. The maps are static, not interactive, although it is possible to zoom and pan. Courtesy of David Rumsey Map Collection, *www.davidrumsey.com*.

Design Considerations

As with the other map types discussed in this chapter, one must ask if the map will add to the act of communication. The principles of good design apply to Web maps just as they do to paper maps or any other graphic, but there are a few additional considerations that result from the medium. Web maps are always viewed on a computer monitor; therefore, one must be aware of the variety of computers and monitors that might be used. There are differences between PC and Macintosh computers and their display capabilities. Laptop monitors and many new desktop monitors are LCD, high resolution, and wide screen. They are 32-bit, meaning they can display over 16 million colors. Older systems may be only 8-bit and lower resolution and can display only 256 colors. Although desktop computers commonly have large monitors, 22 inches or more, "netbooks," supersmall laptop computers, with screens as small as 8 inches, are becoming popular and, increasingly, cell phones are used to access the Web. Thus, since the cartographer has no control over the computer system and monitor, it is best to use Web-safe colors as described in Chapter 4. ColorBrewer indicates if color schemes are optimal on computer monitors.

As with paper maps, color harmony is important for Web maps. Because it is easy to include color backgrounds on Web pages, one should be careful that the map and its background are compatible, and that the lettering is legible on the background and doesn't hurt the user's eyes. The basic guidelines apply, but because of resolution, such things as white lettering on a black background becomes even more difficult to read, and red lettering on an emerald green background seems to vibrate and is annoying to the reader.

Resolution has implications for type, as we saw with animated maps. In general, larger type and type without large differences between thick and thin must be used. Nothing smaller than 10 point should be used. TypeBrewer provides guidelines for the use of type on monitors.

Because monitor sizes vary widely, if a scale is used on the map, it must be a graphic scale. Representative fractions and verbal scales are meaningless when the size of the map changes with the size of the monitor. The guidelines for type also apply, small point sizes cannot be used on the scale. Finally, the scale cannot be broken down into small units. It will usually be a simple reference to show relative sizes and distances, not exact measurements.

In the layout of the "page" it is desirable to keep the map "above the fold" as it is called in newspaper graphic design. This means keeping the map toward the top of the page so that it will not be cut off by some monitors.

It must be remembered that once a map is on the Web, it may be used in ways the designer did not intend; the designer loses control of the map, and since many consider the Web to be public domain, some users will copy it without giving credit.

SUGGESTIONS FOR FURTHER READING

Andrews, Sona Karentz. (1985). Applications of a Cartographic Communication Model to Tactual Map Design. *American Cartographer*, 15, 183–195.

Campbell, Craig S., and Egbert, Stephen. (1990). Animated Mapping: Thirty Years of Scratching the Surface. *Cartographica*, 27(2), 24–46.

Cartwright, William, et al. (Eds.). (1999). *Multimedia Cartography*. Berlin: Springer-Verlag.

Cartwright, William, et al. (Eds.). (2007. *Multimedia Cartography* (2nd ed.). Berlin: Springer-Verlag.

Castner, Henry W. (1983). Tactual Maps and Graphics: Some Implications for Our Study of Visual Cartographic Communication. *Cartographica, 20*(3), 1–16.

DiBiase, David, MacEachren, Alan M., Krygier, John B., and Reves, C. (1992). Animation and the Role of Map Design in Scientific Visualization. *Cartography and Geographic Information Systems, 19*, 201–214, 265–266.

Golledge, Reginald G., and Rice, Matthew. (2005). A Commentary on the Use of Touch for Accessing On-Screen Spatial Representations: The Process of Experiencing Haptic Maps and Graphics. *Professional Geographer, 57*, 339–349.

Harrower, Mark. (2003). Tips for Designing Effective Animated Maps. *Cartographic Perspectives, 44*, 63–65, 82–83.

Harrower, Mark. (2004). A Look at the History and Future of Animated Maps. *Cartographica, 39*(3), 33–42.

Harrower, Mark. (2007). The Cognitive Limits of Animated Maps. *Cartographica, 42*, 349–357.

Kraak, Menno-Jan. (2007). Cartography and the Use of Animation. In William Cartwright, Michael P. Peterson, and Michael P. Gartner (Eds.), *Multimedia Cartography* (pp. 317–326). Berlin: Springer-Verlag.

Kraak, Menno-Jan, and Brown, Allan. (Eds.). (2001). *Web Cartography: Developments and Prospects*. New York: Taylor & Francis.

Krygier, John B. (1994). Sound and Geographic Visualization. In Alan M. MacEachren and D. R. Fraser Taylor (Eds.), *Visualization in Modern Cartography* (pp. 149–166). New York: Pergamon.

Mitchell, Tyler. (2005). *Web Mapping Illustrated*. Sebastapol, CA: O'Reilly.

Olson, Judy M. (1997). Multimedia in Geography: Good, Bad, Ugly, or Cool? *Annals of the Association of American Geographers, 87*(4), 571–578.

PART V

CRITIQUE OF MAPS

Chapter 12

Putting It All Together

> The technological advances that we are experiencing provide the potential for new and better map designs, but without proper cartographic education not only will map quality ultimately suffer but the geographic understanding that is obtained in practicing geographic cartography may be lost as well.
>
> —THOMAS W. HODLER, *Urban
> Geography* (1994)

The final step in creating a map is evaluation and editing. In this chapter we will do "map makeovers" in which we will evaluate maps, analyze their design, and make improvements. Throughout the book we have seen good and bad examples of individual elements, but here we will look at the map as a whole. Anyone who desires to make maps should have a critical eye for maps. The best way to develop this eye is to examine and analyze many maps. The reader is also encouraged to look at maps in newspapers, books, magazines, textbooks, online, and in other media.

It is important that when a map is evaluated that it be critiqued on the basis of the questions listed in Chapter 2; maps are made to serve specific purposes and communicate information to particular audiences, and these goals must be kept in mind. It isn't appropriate to criticize an advertising map for its lack of scientific accuracy, nor is it reasonable to expect a map designed for a college textbook to work equally well in a text for elementary school students. A map shouldn't be made in isolation with no idea of where or how it is intended to be used. This doesn't, of course, mean that the map might not be used for other purposes. The cartographer has no control over its final use, but it should be designed for a perceived audience and purpose.

When evaluating maps, our own or published maps, we must keep in mind the questions that were asked in the beginning and the reality that design is a decision-making process. The questions that must be asked are summarized in Figure 12.1.

Note that in the makeovers that follow the reader may still find problems. Sometimes these are as a result of software capabilities, sometimes they result from shape file problems, some may be differences of opinion. Not all cartographers will agree

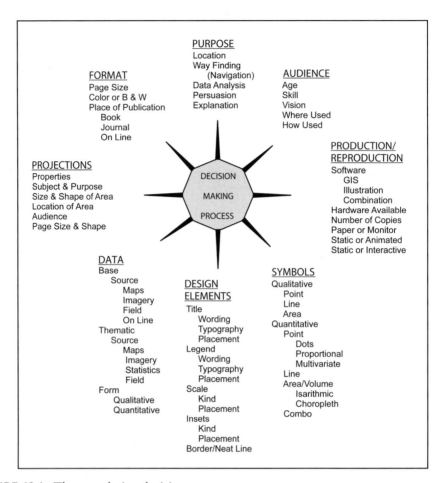

FIGURE 12.1. The map design decision process.

on all aspects of design. Finally, one must keep in mind that there is no perfect map; the cartographer/mapmaker strives to create the best map possible given his or her skills, data, and equipment.

MAKEOVER 1

The cartographer's assignment for Plate 12.1 was to make a map to illustrate an article on poverty in the United States for a popular magazine. The magazine allows color. Plate 12.1 is based on the default colors that are found in some spreadsheet and mapping software programs.

• *Critique.* The colors have no logic; they are simply the additive and subtractive primaries. With this scheme, the lowest value states are red, which would normally be used for high values, and the high-value states are magenta, which can also be used for high.

The map has a large north arrow that serves no purpose and is incorrect for a map with radiating meridians, which can be seen on many north–south state borders. No date is provided and no source of information. While a scale is not strictly necessary on a map of this type, since no measurements will be made, it would be useful for general dimensionality. The categories were selected by the software and have ambiguities. That is, the first category goes up to 9.4 and the second begins with 9.4; therefore, states could be put in either category. In addition, abbreviations are used in the title and legend, which isn't desirable, but can be done if space is an issue.

• *Makeover: Plate 12.2.* The colors have been changed to a sequential scheme with the lightest hues representing the lowest values and the darkest hues representing the highest values. The categories have remained at six, although five or seven could have been used, especially if the mapmaker wanted to show the variation within the highest states or wanted to group the states with less than 10% poverty. A source is given so that the reader can obtain detailed data and a date has been added to the title. A simple scale has been added that gives a sense of relative sizes. Because space is not a problem in the layout, abbreviations have been omitted. No projection name has been provided because there is little distortion on this conic equal-area projection.

MAKEOVER 2

The assignment for the map in Figure 12.2 is to design a map for a college-level U.S. population textbook. The format limits the illustrations to black and white.

• *Critique.* Since this map is designed for a book on the United States, the most important part of the title is the race that is represented. Instead of "Legend," the legend should have a title that elaborates on the subject; it could be "Population in Thousands," which would eliminate the crowding that results from writing out 1,000,000, etc.

However, the two worst errors on this map are (1) using absolute or raw numbers for enumeration areas, which vary widely in size; and (2) using a scale length that is incorrect for the units shown. The length is 1,000 *miles*, not *kilometers*. It would seem that the mapmaker was told to change the scale to kilometers and simply changed units, not length. On this map a scale isn't really needed, but if one is used, it must be correct.

On this map other errors may not be as obvious as on a map in color. The challenge is creating a map with six distinguishable shades of gray.

• *Makeover: Figure 12.3.* First the title was changed to "Asian Population" and the date was added. The data were manipulated to give information as a percent of total population rather than as absolute numbers. Percent gives a better representation of the data because of the widely varying sizes of states and populations. Hawaii, with 39.9% Asian, the highest percentage, does not stand out when raw numbers are used. The word "Legend" was replaced to show the nature of the data. A source statement was added.

On the "before" map, the six shades went from black for the highest to white for the lowest; this isn't necessarily bad, but because Lake Michigan is also shown in

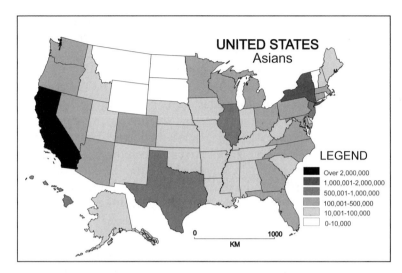

FIGURE 12.2. Asian population map "before."

white, people who don't have a good idea of U.S. geography could assume that it was a state. This shouldn't be a problem with a college textbook. The shades of gray can still be difficult to distinguish depending on the paper used in printing.

MAKEOVER 3

Mike wants a map to advertise his bicycle shop. He will use it in the yellow pages and a newspaper ad. He wants the map to show the location of his shop so that customers can find it easily.

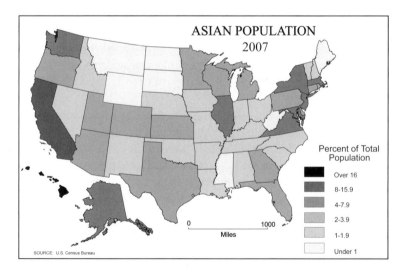

FIGURE 12.3. Asian population map "after."

- *Critique.* The "before" map, Figure 12.4, is not a bad map, but it is inappropriate for its purpose. This map is for advertising, and thus Mike's Bike Shop should stand out. The map will be printed on cheap paper in the newspaper and in the yellow pages so fine lines and small type will not reproduce well.
- *Makeover: Figure 12.5.* The redone map is simpler, but gives address, parking information, and hours of operation, as well as the intersecting streets. Especially now that one can get maps and directions quickly from the Internet, a detailed map isn't needed. The lines and type on the "after" map will reproduce better on cheap paper.

MAKEOVER 4

Suzanne has ridden her bicycle across country, an epic adventure that she wants to recount in a book. The book is a travelogue that will be read by armchair travelers, and especially bicycle tourists who want to make the trip vicariously. Her map will appear on the inside cover of the book if it is hard-bound or as a two-page spread in a paperback. She is restricted to black and white to keep costs down.

- Critique: Suzanne's "before" map, Figure 12.6, is typical of maps that one finds in much travel literature. Surprisingly, many such books contain no maps, so Suzanne's could be considered "better than nothing," but with the space she has available more could be done.

The before map doesn't show the names of the starting and ending points nor the direction of travel. It also has no scale. On this map a scale is necessary to show

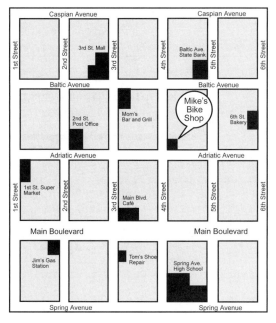

FIGURE 12.4. Bike shop advertising map "before."

FIGURE 12.5. Bike shop advertising map "after."

the length of each leg of the journey. The readers of such books like to follow along on the trip and they want to know how far the traveler went each day and where she stopped. Although Suzanne could put a table at the back of the book listing stops and distances, it is still useful to be able to see the entire route at a glance. Not all of the states are shown on the before map; this isn't a real necessity since most of the target audience will be familiar with the United States.

• *Makeover: Figure 12.7.* For the redone map, the stops along the route are shown and numbered and a legend listing the towns is provided. The starting and ending points are labeled on the map; this also indicates that the trip was from west

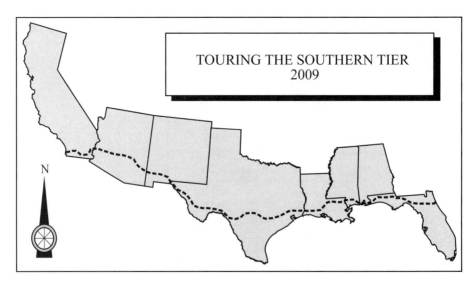

FIGURE 12.6. Bicycle tour map "before."

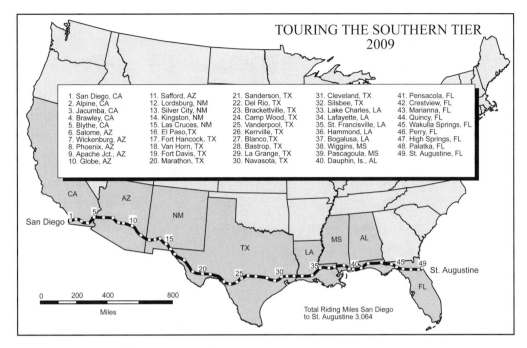

FIGURE 12.7. Bicycle tour map "after."

to east. A scale is provided that allows the reader to determine the approximate riding distance between towns. If Suzanne desires, a table can be provided that shows exact distances. The total distance ridden is also furnished.

MAKEOVER 5

Newspaper maps must often be produced quickly, especially for breaking news, and they must fit a specific format of number of columns and inches. The map in Figure 12.8 is intended to accompany an article on rising unemployment in the business section of a newspaper.

• *Critique.* The "before" map displays the values in an appropriate manner and wouldn't mislead the reader. However, the north arrow is overwhelming, and of course not appropriate. The shades of gray are quite close, especially for the 5–6.4 and 6.5–7.9 categories. Color is permitted in the newspaper, so having a color map would be an attention grabber.

• *Makeover: Plate 12.3.* The redone map is in shades of red, which attract the eye and permit a greater range by adding saturation as well as lightness; this allows the lowest value states to have a tint. The unnecessary north arrow has been removed and an explanatory title has been added to the legend. Although one would hope that readers of the business section would have a knowledge of U.S. geography and know the names and locations of the states, many studies have found that for a large

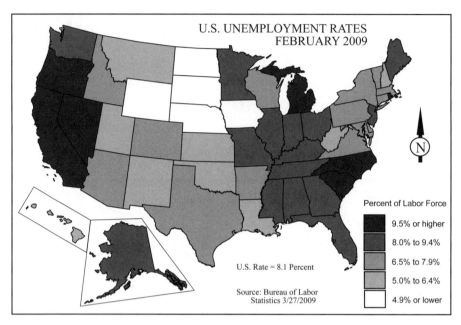

FIGURE 12.8. Newspaper map "before."

percentage of Americans this is not the case. Therefore, postal abbreviations for the states have been added.

MAKEOVER 6

Figure 12.9 is a map designed for a 6th-grade social studies workbook. It is for a section on world ecoregions and shows the combined climates of ice cap and tundra, and tropical and subtropical desert. Each section of the workbook has a map for the subject and a series of questions to aid students in learning to read maps. These questions often involve size and distance comparisons as well as cardinal directions.

• *Critique.* The "before" map is drawn on a Mercator projection. Despite resolutions against the use of cylindrical maps in school textbooks, and the well-known criticisms of the Mercator in particular, a surprising number of school workbooks continue to use rectangular projections, most often the Mercator. In addition, the before map contains an elaborate compass rose that is unnecessary on this projection.

Interestingly, this is one of the most difficult maps to make over because such workbooks must conform to state and national standards and these standards for maps often seem to run counter to good cartographic practice.

• *Makeover: Figure 12.10.* Replacing the Mercator projection is the most obvious improvement. Although it might be desirable to use an equal-area projection, many of these compress the polar regions too much. The Robinson projection was selected because it is a "compromise" projection that has a good appearance and doesn't have

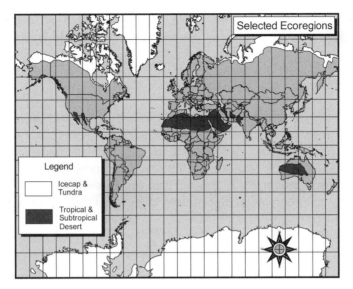

FIGURE 12.9. School workbook map "before."

excessive distortion in any feature—it looks good. It has also become a commonly used world projection in classrooms and school atlases, replacing the Mercator. The compass rose, which, as we have seen, is more appropriate to navigation charts, is a major difficulty. Many of the state standards specifically mention the direction indicator as one of the necessary elements of a map. Maps in workbooks and textbooks include compass roses even on conics or the Robinson, which has curved meridians. Of course, parallels and meridians are the best orientation indicators, but the cartographer could be required to place a compass rose on the map by the publisher. If this

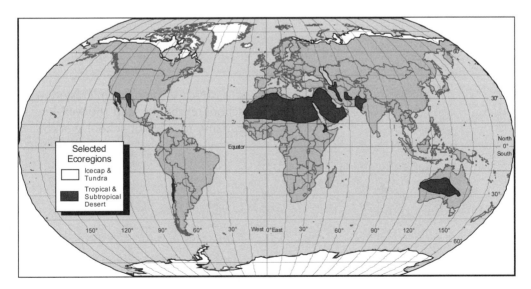

FIGURE 12.10. School workbook map "after."

is absolutely required, the least misleading place to put the rose is where a straight meridian crosses a straight parallel and a less ornate rose could be used.

CONCLUSIONS

For any given set of data, as we have seen throughout this book, there are many possible cartographic representations. The representations vary because of the audience, format, purpose, the symbols chosen, the data manipulation and categorization, the software used, and the map maker's aesthetic sense. Representations also vary based on standards and requirements of agencies and companies that publish maps.

It is hoped that the reader takes away from this book an appreciation of the importance of design as a decision-making process and that he or she will not take that process lightly and simply select default options.

APPENDICES

Appendix A   Commonly Used Projections

MERCATOR

Classification	Cylindrical, conformal.
Graticule	*Meridians:* equally spaced, straight, parallel lines. *Parallels:* unequally spaced, straight, parallel lines; spacing increases toward the poles.
Scale	True along the equator; constant along any parallel, increasing with distance from the equator; infinity at the poles.
Distortion	Increases away from the equator; areal distortion great in high latitudes.
Usage	Designed and recommended for navigation. Recommended and used for conformal maps of equatorial regions.
Other	All rhumb lines are straight lines.

MILLER CYLINDRICAL

Classification	Cylindrical, neither equal area nor conformal.
Graticule	*Meridians:* equally spaced, straight, parallel lines. *Parallels:* unequally spaced, straight, parallel lines; closest at the equator.
Scale	True along the equator; constant along any parallel, changes with latitude and direction.
Distortion	Increases away from the equator.
Usage	World maps.

SINUSOIDAL

Classification	Pseudocylindrical; equal area.
Graticule	*Meridians:* central meridian is a straight line one-half as long as the equator; others are equally spaced sine curves. *Parallels:* equally spaced, straight, parallel lines; perpendicular to the central meridian.
Scale	True on the parallels and central meridian.
Distortion	Accurate along equator and central meridian.
Usage	Atlas maps of South America and Africa; occasionally for world maps.
Other	Also called Sanson–Flamsteed projection.

MOLLWEIDE

Classification	Pseudocylindrical; equal area.

A poster-size chart of projections is available from USGS at *egsc.usgs.gov/isb/pubs/MapProjections/projections/html*

Graticule	*Meridians:* central meridian is a straight line one-half as long as the equator. Meridians 90° east and west of the central meridian form a circle; others form semiellipses. *Parallels:* unequally spaced, straight, parallel lines, perpendicular to the central meridian.
Scale	True along 40°N and S; constant along any given latitude.
Distortion	Severe near the outer meridians at high latitudes.
Usage	World atlas maps, especially thematic maps.
Other	Also called homolographic; often used in interrupted form.

ECKERT IV

Classification	Pseudocylindrical; equal area.
Graticule	*Meridians:* central meridian is a straight line one-half the length of the equator; others are equally spaced semiellipses. *Parallels:* unequally spaced, straight, parallel lines; perpendicular to the central meridian; widest spacing is near the equator. Poles are lines one-half the length of the equator.
Scale	True along 40°30′; constant along any given latitude.
Distortion	Distortion-free at 40°30′N and S at central meridian.
Usage	Thematic and other world maps in atlases and textbooks.

GOODE'S HOMOLOSINE

Classification	Pseudocylindrical; composite; equal area.
Graticule	*Meridians:* in interrupted form, six straight central meridians; kink in meridians at 40°44′N and S. *Parallels:* straight parallel lines perpendicular to the central meridians. Poles are points.
Scale	True along every latitude between 40°44′ N and S and along central meridian between 40°N and S.
Distortion	Same as sinusoidal between 40°44′N and S; same as Mollweide beyond this range.
Usage	Almost always presented in interrupted form; numerous world maps in atlases and textbooks.
Other	Composite of Mollweide (homolographic) and sinusoidal.

ROBINSON

Classification	Pseudocylindrical; neither equal area nor conformal.
Graticule	*Meridians:* central meridian is a straight line; others resemble elliptical arcs. *Parallels:* straight, parallel lines; equally spaced between 38°N and S; spacing decreases beyond these limits.
Scale	True along 38°N and S; constant along any given latitude.
Distortion	Low within 45° of the central meridian and along the equator; no point is free of distortion.
Usage	Thematic world maps.

LAMBERT CONFORMAL CONIC

Classification	Conic; conformal.

Graticule	*Meridians:* equally spaced, straight lines converging at a common point (one of the poles). *Parallels:* unequally spaced, concentric circle arcs centered on the pole of convergence.
Scale	True along one or two chosen standard parallels; constant along any given parallel.
Distortion	Free of distance distortion along the one or two standard parallels.
Usage	Midlatitude regions of E–W extent.
Other	Cannot show entire earth.

ALBERS EQUAL-AREA CONIC

Classification	Conic; equal area.
Graticule	*Meridians:* equally spaced, straight lines converging at a common point, normally beyond the pole. *Parallels:* unequally spaced concentric arcs. Poles are circular arcs.
Scale	True along the one or two standard parallels; constant along any given parallel.
Distortion	Free of angular and scale distortion only along the one or two standard parallels.
Usage	Frequently used for U.S. maps, thematic maps; recommended for equal-area maps of E–W midlatitude regions.

POLYCONIC

Classification	Polyconic; neither equal area nor conformal.
Graticule	*Meridians:* central meridian is a straight line; all others are complex curves. *Parallels:* equator is a straight line; all others are nonconcentric circular arcs spaced at true distances along the central meridian.
Scale	True along the central meridian and each parallel.
Distortion	Distortion-free only along the central meridian; extensive distortion if there is a great E–W range.
Usage	Sole projection for USGS topographic maps until the 1950s.
Other	Not recommended for regional maps.

GNOMONIC

Classification	Azimuthal; perspective; neither equal area nor conformal.
Graticule	Polar aspect. *Meridians:* equally spaced, straight lines intersecting at the central pole. *Parallels:* unequally spaced, concentric circles centered on the pole; spacing increases toward the equator. Equator and opposite pole cannot be shown.
Scale	True only at the center.
Distortion	Only the center is distortion-free; distortion increases rapidly away from the center.
Usage	To show great circle paths as straight lines; for navigation.
Other	All great circle arcs show as straight lines; can be used in polar, equatorial, and oblique aspects.

STEREOGRAPHIC

Classification	Azimuthal; conformal; perspective.
Graticule	Polar aspect. *Meridians:* equally spaced, straight lines intersecting at the central pole. *Parallels:* unequally spaced, concentric circles centered on the pole. Opposite pole cannot be shown.
Scale	True only at center.
Distortion	Only the center is distortion-free.
Usage	For topographic maps of polar regions.
Other	All great or small circles show as circles or straight lines. Can be used in polar, equatorial, or oblique aspects.

ORTHOGRAPHIC

Classification	Azimuthal; perspective; neither conformal nor equal area.
Graticule	Polar aspect. *Meridians:* equally spaced, straight lines intersecting at the central pole. *Parallels:* unequally spaced circles centered at the pole; spacing decreases away from the central pole. Only one hemisphere can be shown.
Scale	True only at the center.
Distortion	Only the center is distortion-free.
Usage	Pictorial views of the earth and the moon.
Other	Has the look of a globe. Can be used in polar, equatorial, or oblique aspects.

AZIMUTHAL EQUIDISTANT

Classification	Azimuthal; equidistant.
Graticule	Polar aspect. *Meridians:* equally spaced, straight lines intersecting at the central pole. *Parallels:* equally spaced circles centered at the pole. The entire can be shown.
Scale	True along any straight line radiating from the center.
Distortion	Only the center is distortion-free.
Usage	Polar aspect is used for maps of the polar region; oblique aspect is frequently centered on major cities to show distances.
Other	Also called zenithal equidistant. Can be used in polar, equatorial, or oblique aspects.

AZIMUTHAL EQUAL-AREA

Classification	Azimuthal; equal area.
Graticule	Polar aspect. *Meridians:* equally spaced straight lines intersecting at the central pole. *Parallels:* unequally spaced circles; spacing decreases away from the pole. Entire earth can be shown.
Scale	True only at the center.
Distortion	Only the center is distortion-free.
Usage	Polar aspect is frequently used for atlas maps of polar regions.
Other	Can be used in polar, equatorial, or oblique aspects.

Appendix B Resources

A major part of knowing where to look is knowing what to look for.

—MARK MONMONIER, *Mapping It Out* (1993)

CARTOGRAPHIC ORGANIZATIONS

Association of American Geographers, Cartography Specialty Group
www.csun.edu~hfgeg003/csg

British Cartographic Society
www.cartography.org.uk
Journal: *The Cartographic Journal*

Canadian Cartographic Association
www.cca-acc.org
Journal: *Cartographica*

Cartography and Geographic Information Systems Society (CaGIS)
www.cagis.org
Journal: *CaGIS Journal*

International Cartographic Association (ICA)
cartography.tuwein.ac.at/ica

North American Cartographic Information Society (NACIS)
www.nacis.org
Journal: *Cartographic Perspectives*

REGIONAL MAP SOCIETIES

California Map Society
www.californiamapsociety.org/index.html

Chicago Map Society
www.newberry.org/smith/cms/cms.html

New York Map Society
www.nymapsociety.org

Texas Map Society
libraries.uta.edu/txmapsociety

Washington Map Society
home.earthlink.net/~docktor/washmap.htm
Journal: *The Portolan*

TOOLS

Color

ColorBrewer
  www.colorbrewer.org

Cartograms

Cartogram Central
  www.ncgia.ucsb.edu/Cartogram_Central/index.html

World Mapper
  www.worldmapper.org

Type

TypeBrewer
  www.typebrewer.org

Map Projections

www.csiss.org/map-projections/index.html

www.nationalatlas.gov/articles/mapping/a_projections.html

mapthematics.com/Projections

General

ESRI Mapping Center
  mappingcenter.esri.com

U.S. GOVERNMENT AGENCIES

Bureau of the Census
  www.census.gov

National Aeronautics and Space Administration (NASA)
  www.nasa.gove/home/indes/index.html

National Atlas
  www.nationalatlas.gov

National Geospatial-Intellegence Agency (NGA)
  www1.nga.mil/pages/default.aspn

National Oceanic and Atmospheric Administration (NOAA)
  www.noaa.gov

U.S. Board of Geographical Names (a part of USGS)
  geonames.usgs.gov/index.html

United States Geological Survey (USGS)
  www.usgs.gov/pubprod

SOFTWARE COMPANIES

Adobe Illustrator
  www.adobe.com/products/illustrator

Atlas GIS
  rpmconsulting.com/atlas/software.html

CorelDraw
  www.corel.com

ESRI
  www.esri.com

MapInfo
  www.mapinfo.com

MAPublisher
  www.avenza.com/products.mapub.html

Ortelius
  www.mapdiva.com

ONLINE MAPPING

Google Earth
  earth.google.com

Google Maps
  maps.google.com/

MapQuest
  www.mapquest.com

Navteq
  www.navteq.com

MAPPING COMPANIES AND ORGANIZATIONS

Adventure Cycling
  www.adventurecycling.org

DeLorme Maps (including career information)
  www.delorme.com/about/default.aspx

Hammond
  www.hammondmap.com/catalog/index.php

National Geographic Society
  www.nationalgeographic.com

Rand McNally
  www.randmcnally.com

Raven Maps (Allen Maps)
  www.ravenmaps.com
  www.allancartography.com

Thomas Brothers RandMcNally
  www.mapbooks4u.com

FORUMS AND OTHER LINKS

Cartographic Users Advisory Council
  cuac.wustl.edu

Cartography On Line
  CartographyOnline.com/index.php

CartoTalk (Cartography and design forum)
  www.cartotalk.com

Oddens Bookmarks
  oddens.geog.uu.nl/index.php

HISTORY OF CARTOGRAPHY

David Rumsey (Historic map collection online)
  www.davidrumsey.com

Herman Dunlop Smith Center for the History of Cartography
  www.newberry.org/smith/smithhome.html
  Newsletter: *Mapline*

Imago Mundi (Journal for history of cartography)
  www.maphistory.info/imago.html

Appendix C Glossary

> "When I use a word," Humpty Dumpty said, in rather a scornful tone, "it means just what I choose it to mean—neither more nor less."
>
> —LEWIS Carroll, *Through the Looking Glass* (1896)

These definitions are the generally accepted meanings of the terms used and are intended to provide the reader with a working knowledge of the terms.

Abstract symbol. Also called "arbitrary symbol." A type of symbol that is not visually or conceptually related to the object represented. Such symbols are frequently geometric shapes, such as circles, squares, and triangles.

Additive primaries. Three basic colors of light from which other hues can be produced; they are red, green, and blue. Contrasts with *subtractive primaries*.

Altitude tint. *See* Layer tint.

Angular deformation. Alteration in angles and shape caused by the transformation of the sphere on a plane surface.

Animated maps. Maps that create the illusion of change, either temporal or spatial, by rapidly displaying a series of single frames (Peterson, 1995).

Area of least deformation. The area on a projection that is most accurate. Also called the "zone of best representation."

Areal phenomena. Geographical phenomena that extend over defined areas. These phenomena have two significant dimensions: length and width.

Associative symbol. A type of symbol midway between abstract and pictorial. Although associative symbols are not small pictures of the object represented, they give an impression of the nature of the object in some way. Thus, a mountain peak might be represented by a triangle or a building by a small square or rectangle. *See also* Pictorial symbol; Mimetic symbol; Abstract symbol.

Atlas. A collection of maps, bound or boxed, that conform to a uniform format.

Azimuth (magnetic). A spherical angle formed by a true north line (a meridian) and a line that passes through the observer and the object observed. The spherical angle between any great circle and a meridian.

Azimuthal projections. A class of projections that shows azimuths correctly from the center point; great circles through the center point are straight lines. Also called "zenithal projections."

Balance, visual. The positioning of various graphic elements so that the visual weight of the various elements is equally distributed.

Bar graph. A type of graph that shows value by means of bars whose height is proportional to the value represented.

Bar scale. *See* graphic scale.

Base data. Fundamental information such as coastlines, boundaries, and the like that provides a framework for thematic data.

Base line. (1) A surveyed line to which other surveys are referred. (2) In typography, the line on which letters are arranged.

Base map. A map used as a primary source for compilation (*see* Source map) or as a framework on which new detail in printed.

Binary color schemes. A scheme that uses only two hues or one hue plus lightness to represent "either–or" situations, such as rural/urban, forest/grassland, and the like.

Bivariate cartograms. A value-by-area cartogram on which one variable is shown by the cartogram and another variable by shading or coloring the areas.

Bivariate choropleth. A choropleth map that shows two related variables.

Bleeding. A bleeding edge is an edge of a map to which printed detail extends after the paper has been trimmed. Extending the mapped area through the normal border is also called "bleeding the map."

Cadastral map. A map that delineates property boundaries.

Callout. A label that is not adjacent to its symbol, but has a line or arrow pointing to the symbol. Often the label is enclosed in a frame.

Cartogram. An abstracted and simplified map for which the base is not true to geographic scale. In the most common forms, the areas of features are drawn according to a value, such as population, or a time scale is used instead of a distance scale. *See also* Linear cartogram and Value-by-area cartogram.

Cartography. The art, science, and technology of making maps, together with their study as scientific documents and works of art.

Center point cartogram. A cartogram that shows distance on a time scale from a center point.

Chernoff faces. A symbol that shows multivariate data as a human face with individual features, eyes, nose, mouth, and the like, representing the values of the variables. Introduced in 1973 by Herman Chernoff.

Choropleth map. A type of quantitative map on which statistical or administrative areas are colored or shaded proportionally to the value represented.

Chroma. *See* Saturation.

Clarity. In map design this refers to ease of understanding the map. On a clear map all elements serve a purpose and there is no distracting clutter or "noise."

Classification. Placing data into groups that have similar characteristics or values. A generalization process.

Classless choropleth. A type of choropleth map for which the data are not placed into categories. Instead, interval scaling of the data is employed, and each enumeration area is represented by a separate areal symbol, usually crosshatched lines. Also called "unclassed choropleth" or "cross-line choropleth."

CMYK. The subtractive primary colors, cyan, magenta, and yellow, plus black. These are the primaries used with pigments on paper.

Color separation. Separation of the colors on a map by photographic or drafting processes for subsequent reproduction. A separate drawing, negative, or scribed image is prepared for each color.

Communication. Transferring a message through some medium to a receiver.

Compilation. Selection, assembly, evaluation, and graphic presentation of all relevant information required for the preparation of a map. Such information may be derived from other maps or from other sources.

Composition. The arrangement within the borders of map elements, such as subject area, title, legend, scale, and the like.

Compromise projection. A projection that has no special properties, but usually has a good appearance.

Condensed projections. Projections having a portion of the grid removed to permit a larger scale map to be presented on a given page size.

Conformal projections. Projections on which the shapes of very small areas are preserved. Parallels and meridians cross at right angles and scale is the same about a point.

Conic projections. Projections that appear to have been projected onto a cone.

Contiguous cartogram. A value-by-area cartogram in which the different areas represented preserve boundaries. Adjacent areas retain internal boundaries.

Continuous distribution. A geographic distribution that occurs everywhere within the mapped area.

Contour interval. The vertical distance between two adjacent contours.

Contour line. A line that joins all points having the same elevation above or below a datum, usually mean sea level. A type of isarithmic line, also called an "isohypse."

Contour spacing. The horizontal or map distance between two adjacent contour lines.

Contrast. Differences in light and dark shades, thick and thin lines, large or small type, rough or smooth textures or patterns on a map. A map that lacks contrast is visually uninteresting and hard to read.

Control points. Points on a map that have known horizontal locations and known values that are used in estimating positions for isarithmic lines. Control points for contour lines are spot elevations; for isotherms the points may be weather stations.

Conventional projection. A term used in some classifications of projections to describe all except azimuthal, cylindrical, and conic projections. Also called "mathematical" projections.

Cybercartography. A term coined by D. R. F. Taylor in 1997 to describe mapping on the Web and virtual cartography.

Cylindrical projections. Projections that appear to have been created by projecting the earth's grid onto a cylinder.

Dasymetric map. A variation of the choropleth map. Color or shading is applied to areas of homogeneity and is not restricted to administrative or statistical boundaries.

Data classification. Grouping of data into categories or numerical classes.

Data model. A conceptual, three-dimensional model in which blocks or prisms are constructed vertically from each statistical or administrative area in proportion to its actual or mapped value. It is used in visualizing the statistical surface for choroplethic maps.

Datum. Any value that serves as a reference or base. For contour maps, the datum is commonly sea level.

Defined scale. *See* Nominal scale.

Demers cartogram. Similar to Dorling cartograms. Enumeration areas are replaced by squares that are proportional to the value represented.

Design. (1) The process of creating a map. (2) A plan of execution. (3) The appearance of a map.

Developable surface. A surface that can be cut and flattened without distortion. The three developable surfaces are the cylinder, the cone, and the plane.

Direction. The location of one point in space with respect to another.

Discontinuous distribution. A distribution that does not occur everywhere in the mapped area.

Distance-by-time cartogram. A type of cartogram that uses time instead of linear distance as a scale.

Diverging color scheme. A color scheme used to represent increases and decreases of a variable from a midpoint, such as temperatures above and below freezing or positive and negative change.

Dorling cartogram. A value-by-area cartogram on which the enumeration areas are replaced by uniform geometric figures, usually circles, which are drawn proportional to the value represented.

Dot density map. A variation of the dot map that places dots randomly within the enumeration area.

Dot map. A representation of geographic phenomena on which dots represent a specified number of the phenomenon being mapped. The dots are placed in their area of occurrence within the enumeration area.

Dynamic symbols. Static symbols that give an impression of action or movement, such as bomb bursts or flames.

Earth's grid. The system of parallels and meridians on the earth or a globe. Also called the "graticule."

Electromagnetic spectrum. The ordered array of known electromagnetic radiations from cosmic rays at the short-wavelength end of the spectrum to radio waves at the long-wavelength end of the spectrum. Electromagnetic energy is all energy that moves with the speed of light.

Equal-area projection. A map projection that preserves a uniform area scale. Also called an "equivalent" projection.

Equator. An imaginary great circle drawn around the surface of the earth midway between the north and south poles. It divides the earth into two equal hemispheres and is designated as 0° latitude.

Equidistant projections. Projections that show distance correctly along certain lines or from certain points.

Equivalent projections. Projections on which countries, continents, and other areas maintain their correct areal scale. Also called "equal-area" projections.

Exaggeration. An operation of generalization in which features are made larger.

External base data. These include title, legend, scale, north arrow, grid, and text. The external base data have an explanatory function.

Figure–ground relationship. The relationship between the main shape or figure in a graphic display and the background.

Flow line. A linear symbol whose width varies in direct proportion to the quantity represented.

Font. All the letters, numbers, special characters, and variations of a typeface.

Format. The size and shape of a map.

Generalization. A basic cartographic procedure that reduces the amount of information presented in order to create a clearer communication. Because maps are drawn smaller than reality, they must be generalized. The processes involved in generalization are selection, simplification, classification, and symbolization.

General purpose map. A map that shows a variety of geographical phenomena, such as transportation, political boundaries, hydrography, and the like. It is used primarily for reference, location, and planning.

Generating globe. A model of the earth used for the generation of perspective map projections, or a theoretical sphere to which projections may be compared. The scale of the generating globe is the nominal scale of the map.

Geographic data. Facts gathered by measuring, counting, calculating, or derivation. They measure or describe aspects of geographical phenomena.

Geographic information system. A computer-based system for collecting, managing, analyzing, modeling, and presenting geographic data for a wide range of applications.

Geoid. An earth-shaped figure; the figure of the earth viewed as a mean sea-level surface.

Geometric symbol. An abstract symbol that uses geometric shapes such as triangles, squares, and cubes.

GIScience. This term is used in two ways. (1) As a field of research that studies theory and concepts of GIS. (2) Applying scientific method to address spatial questions. GIScientists, in this sense, practice or use GIS. GIS technician, analyst, or specialist is often used as a synonym for GIScientist.

Gnomonic projection. An azimuthal projection that shows all great circles as straight lines.

Graduated circle. A circular symbol that is drawn so that its area or apparent area is proportional to the amount represented: also called a "proportional circle."

Graphic (or graphical) scale. A graduated line marked in ground units that allows distances to be measured from a map. Also called a "bar scale."

Graticule. The system of parallels and meridians on the earth or globe.

Gray scale. Standard set of gray tones ranging from white to black.

Great circle. A circle on the earth's surface formed by a plane passed through the earth's center that bisects it into equal hemispheres. The shortest distance between two points of the earth's surface is along a great circle arc. Also called an "orthodrome."

Grid ticks. Short lines indicating where selected grid lines intersect the neat line.

Halos. Light outlines around letters.

Haptic. Pertaining to touch. Haptic maps use force-feedback devices that allow the user to feel map features.

Harmony. One of the goals of graphic design. Harmony refers to all elements of the map being agreeably related.

Hill shading. Shading used to create a three-dimensional impression of relief.

Histogram. A graph with bars in which the area of the bars is proportional to the frequency in each class or group.

Hue. The dimension of color related to its wavelength. This is the main characteristic of a color and distinguishes it from other colors.

Hypsometric tint. Color applied to the area between two selected contours. It is a method of showing relief on maps. Commonly, cool colors are used for low elevations and warm colors for high elevations. *See also* Layer tint.

Imagery. Visible representation of objects from cameras, scanners, radar, or other sensors.

Inset map. A separate map located within the neat line of a larger map. It may be a portion of the map at an enlarged scale, an area that falls outside the borders but included within for convenience, or a smaller scale map of surrounding areas included for location purposes.

Intensity. The richness of a color. Other terms used are "chroma," "saturation," and "purity."

Interactive Map. A map that allows the user to click on parts of the map and interact with various features.

Internal base data. These data, which are a part of the base map, include such things as administrative boundaries, coastlines, cities, transportation routes, place names, mountains, rivers, and lakes, that is, geographic information on or within the mapped area itself.

Interpolation. Estimating values from an isarithmic map; estimating location of isarithmic lines from control points.

Interrupted projections. Projections on which the central meridian is repeated in order to reduce distortion. It permits each area (usually a continent) to be placed in the zone of best representation. Also called a "recentered projection."

Interval, choroplethic and isarithmic. The difference in value between two adjacent categories or isarithmic lines.

Interval measurement, or interval scale. Provides information about differences in value between ranks, in addition to rank and kind. *See also* Nominal measurement and Ordinal measurement.

Isarithm. A line that joins all points having the same value above or below some datum. Also called "isoline" or "isogram."

Isoanamorphic lines. Lines that show equal deformation on a map projection.

Isochrone. A line that joins all points having the same time from a specific point.

Isoline. Another term for "isarithm."

Isometric line. A type of isarithmic line. The lines connect values either derived or actual that occur at actual points.

Isopleth. A type of isarithmic line. The lines connect derived values that occur at inferred points. The lines are assumed to have equal values for the distribution.

Key. Synonym for *legend*. Often used on children's maps.

Latitude. Angular distance north or south of the equator; measured from 0° to 90°.

Layer tint. The use of colors between different isarithmic lines. When used with elevation contours, the method is called "hypsometric tints."

Layout. The arrangement of the various map elements within the borders of the map. *See also* Composition.

Legend. An explanation of the symbols and conventions used on a map.

Lightness. Also called "value." Shades of a hue from light to dark.

Linear cartogram. Cartogram that is concerned with distances.

Linear phenomena. Geographic phenomena that are linear in form, such as roads, railroads, rivers, and boundaries. Linear phenomena have only one significant dimension: length.

Linear scale. *See* Graphic scale.

Linear symbol. A line used to represent geographic phenomena that are linear in nature.

Longitude. Angular distance east or west of the prime meridian; measures from 0° to 180°.

Loxodrome. *See* Rhumb line.

Map. A graphic representation of all or a part of the surface of the earth (or other body) drawn to scale on a plane.

Menu. A list of options on a display screen or digitizing tablet that allows the computer operator to choose procedures.

Meridian. A line that connects all points having the same longitude. Meridians are great circles and converge at the poles.

Mimetic symbol. A symbol that is more or less visually imitative of the object represented, such as an airplane to represent airports. *See* Pictorial symbol.

Modern typeface. A type style with a vertical and symmetrical stress to the letters with a sharp distinction between thick and thin strokes. The term "modern" does not indicate the age of any given typeface. *See also* Old Style typeface.

Natural scale. A synonym for "representative fraction."

Nautical mile. A unit of distance equal to 1 minute of arc along a great circle or 1,852 meters or 6,076.1 feet.

Neat line. A line bounding the detail of the map.

Noise. Anything that interferes with the map communication.

Nominal measurement. Distinguishes between things or classifies them on the basis of their nature without any indication of quantity.

Nominal scale. The scale used on the map. It is correct only for certain lines or from certain points.

Noncontiguous cartogram. A type of value-by-area cartogram on which the areas represented are separate from one another. Internal boundaries are not retained.

Oldstyle typeface. A type style that has a diagonal stress with asymmetric curves and subtle differences between thick and thin strokes. The term "Old Style" does not indicate the age or time when a face was designed. *See also* Modern typeface.

Optical center. The apparent center of a page or layout. It is slightly above the geometric center.

Order. Arrangement of graphic elements in a logical manner.

Ordinal measurement. Classifies and ranks data without specifying numerical values.

Orthodrome. *See* Great circle.

Orthogonal. At right angles.

Orthomorphic. A synonym for "conformal."

Parallel. A line that joins all points having the same latitude. Parallels are true east–west lines that encircle the globe. Only the equator is a great circle.

Perspective projection. A projection that can be developed geometrically or by a light source from a generating globe.

Persuasive map. A type of map whose main object or effect is to change or in some way influence the reader's opinion. Advertising maps and propaganda maps are both examples of persuasive maps.

Photogrammetry. The science of obtaining accurate measurements from photographs; mapping or surveying by photographic methods.

Pictorial symbol. A symbol on a map that is a small picture of some object. Normally, these symbols imitate the object or distribution being mapped. Also called "mimetic symbols" or "pictographs."

Pie chart. A circular symbol divided into sectors to indicate proportions of a total value. Often combined with proportional circles. Also called "pie graph," "segmented circle," or "sectored circle."

Plane projection. A transformation of the earth's grid onto a plane surface.

Planimetric. Features are shown in their correct horizontal positions.

Point phenomena. Geographic phenomena that occur or are assumed to occur at discrete points.

Polygonal graph. Similar to a radar graph, but the rays are not shown.

Primaries. The three colors used to create all other colors. The additive primaries are red, green, and blue; the subtractive primaries are cyan, magenta, yellow, and black.

Prime meridian. The meridian adopted at the origin (0°) for measurement of longitude. The meridian passing through Greenwich, England, was adopted in 1884, and is accepted by most countries of the world. Prior to that date, each country chose its own prime meridian, frequently the meridian through its capital city.

Printer. (1) The person who prints a map. (2) A piece of computer hardware that is capable of printing letters and perhaps graphics.

Profile. A cross section of a part of the earth's surface created by plotting elevations from a contour map along a linear traverse. The horizontal scale represents linear distance along the traverse, and the vertical scale represents elevation. The vertical scale is commonly exaggerated to bring out variations in the terrain.

Projection. A systematic arrangement of all or a part of the earth's (or other spherical) grid on a plane.

Propaganda map. A map designed to persuade or influence the reader; the connotation of propaganda is usually negative or untruthful.

Proportional symbol. A point symbol, such as a circle or square, that is drawn so that its area is actually or visually proportional to the amount represented.

Pseudocylindrical projection. A projection on which parallels are straight lines of varying length and meridians are curved lines that are spaced equally on the parallels.

Purity. *See* Saturation.

Qualitative symbols. Symbols that show some nonquantitative aspect of a geographic phenomenon. Symbols that represent location of nominally scaled data.

Quantitative symbols. Symbols that show some quantitative aspect of a geographic phenomenon.

Radar graph. A multivariate point graph that shows variables as rays.

Range graded. Grouping of data into similar categories.

Ratio measurement. Classifies and gives differences between values using a scale that starts at absolute zero.

Rectangular cartogram. A variety of value-by-area cartogram on which the different areas are represented by squares, rectangles, and other geometric figures with little or no attempt to approximate the true shapes of the areas.

Reference globe. *See* Generating globe.

Registration. The alignment of multiple images, such as the various flaps used to prepare a color map.

Remote sensing. Detection and/or recording of data about an object without being in physical contact with the object.

Representative fraction. The scale of a map expressed as a fraction or ratio that relates distance on the map to distance on the ground in the same units. Also called "natural scale" or "fractional scale."

RGB. The additive primary colors: red, green, and blue.

Rhumb line. A line that cuts all meridians at a constant angle. Also called a "line of constant compass direction," "line of constant bearing," or "loxodrome."

Sans serif. Type styles that do not have serifs.

Saturation. The "colorfulness" or vividness of a color. The closer a hue is to gray, the less saturated it is. Also called "chroma."

Scale. The ratio of the distances on a map, globe, model, or profile to the actual distances they represent.

Scale factor. The ratio of the scale of a projection to that of the generating globe. The scale factor at standard points or along standard lines is 1.0.

Segmented proportional circles. *See* Pie charts.

Selection. A generalization process. Selection involves choosing the information that will be shown on the map.

Semimimetic symbols. Symbols that fall between pictorial and abstract. While not pictorial, they give the "feel" of the object symbolized, such as a triangle to represent a mountain peak.

Semiotics. The theory of signs and symbols.

Sequential color scheme. One that uses one or two hues with lightness steps to illustrate a single variable.

Serif. The short cross lines found at the ends of the main strokes of letters in some typefaces.

Simplification. A generalization process. It involves elimination of unnecessary detail.

Simultaneous contrast. An effect that causes a color to appear tinged with the complement of its surrounding color. A special case of simultaneous contrast called "induction" causes adjacent colors on a scale to appear lighter next to darker colors and darker next to lighter colors.

Slope. The inclination of a surface with respect to the horizontal.

Small circle. A circle on the earth's surface whose plane does not pass through the earth's center. *See also* Great circle.

Smoothing. A generalization process that eliminates some detail of complex linear features, such as coastlines.

Snowflake graph. Similar to a radar graph except the rays are joined to form a polygon.

Software. Programs and data files for a computer. Programs provide instructions to the computer, data files provide information.

Sound maps. Maps that use sound as a symbol.

Source map. A map used as a base or source in compilation.

Spatial data. Information for geographic or spatial phenomena.

Spatial phenomena. Features in the real world that have a spatial component.

Spectral color scheme. A color scheme that uses the colors of the visible spectrum as steps.

Spheroid. A geometric figure describing the size and shape of the earth, the departures from true sphericity being determined from measurements of the earth's surface.

Standard parallel. A parallel on a conic projection that has no distortion. It is the line where the assumed developable surface touches the generating globe.

Standard points and standard lines. Points and lines on a projection that have no distortion.

Star graph. *See* Snowflake graph.

Statistical surface. A conceptual three-dimensional surface formed by horizontal location on the xy plane and statistical values in the vertical, z, dimension. *See also* z values.

Statute mile. A unit of distance used in the United States and Great Britain. It is equal to 5,280 feet or 1.609 kilometers.

Subtractive primaries. Three basic pigments from which other hues can be produced. The subtractive primaries, magenta, cyan, and yellow, are used with printing inks and paints. *See also* Additive primaries.

Symbol. A mark placed on a map that by convention, usage, or reference to a legend is understood to represent a specific characteristic or feature.

Symbolization. The process of designing or selecting symbols for a map.

Tactile map. A map used by the visually impaired on which all symbols are raised and lettering is in Braille.

Thematic map. A map that features a single distribution, concept, or relationship and for which the base data serve only as a framework to locate the distribution being mapped.

Trivariate choropleth map. A choropleth map that shows three related variables.

Type. (1) The shape of the letter image. (2) The actual piece of metal or photographic image from which the printed image is derived.

Typeface. A type style or design.

Typography. The style of type or the arrangement or appearance of type on the printed page.

Unclassed choropleth. A type of choropleth map for which the data are not placed into categories. Instead, interval scaling of the data is employed, and each enumeration area is represented by a separate areal symbol, usually crosshatched lines. Also called "classless choropleth" or "cross-line cholorpleth."

Unity. A goal of graphic design. The impression that all elements are integrated.

Value. (1) A quantity to be represented. (2) A dimension of color referring to lightness or darkness.

Value-by-area cartogram. A cartogram on which the size of the different units is proportional to some variable, such as population or income, not to actual geographic size.

Verbal scale. A statement in words of the map scale, such as one inch to the mile, or one centimeter to the kilometer.

Visible spectrum. That part of the electromagnetic spectrum that is visible to the human eye. It extends from violet at the short-wavelength end through indigo, blue, green, yellow, and orange, to red at the long-wavelength end.

Visual hierarchy. Refers to different levels of emphasis of the elements of a graphic design.

Visual variables of symbols. Graphic characteristics of symbols. These include size, shape, tonal value, hue, orientation, and pattern.

Visualization. Visualization involves exploration of data and seeing it in different ways. It is often associated with dynamic visual displays. The goal is insight into the data. Some animated maps such as "flythroughs" and "flybys" are also called visualizations.

Volume phenomena. Geographical phenomena that extend over areas but can be conceived of as having a third dimension. The third dimension is a value or quantity. *See also* Statistical surface.

White light. Light that is made up of all colors of the spectrum.

White space. The portion of the page not taken up by the subject.

Z values. On statistical maps, the z values are the values associated with points or areas.

Zenithal projection. A term used synonymously with azimuthal projection.

Zone of least deformation. The area on a projection that is most accurate. Sometimes called the " area or zone of best representation."

Bibliography

American Cartographic Association Committee on Map Projections. (1989). Geographers and Cartographers Urge End to Popular Use of Rectangular Maps. *American Cartographer, 16*, 222–223.

Andrews, Sona Karentz. (1985). Applications of a Cartographic Communication Model to Tactual Map Design. *American Cartographer, 15*, 183–195.

Arnheim, Rudolf. (1969). *Art and Visual Perception: A Psychology of the Creative Eye.* Berkeley and Los Angeles: University of California Press.

Arnheim, Rudolf. (1971). *Visual Thinking.* Berkeley and Los Angeles: University of California Press.

Aronoff, Stan. (2005). *Remote Sensing for GIS Managers.* Redlands, CA: Esri Press.

Bertin, Jacques. (1983). *Semiology of Graphics* (William J. Berg, Trans.). Madison: University of Wisconsin Press.

Bielstein, Susan M. (2006). *Permissions, a Survival Guide: Blunt Talk about Art as Intellectual Property.* Chicago: University of Chicago Press.

Brewer, Cynthia A. (1994). Color Use Guidelines for Mapping and Visualization. In Alan M. MacEachren and D. R. Fraser Taylor (Eds.), *Visualization in Modern Cartography* (pp. 123–147). New York: Pergamon.

Brewer, Cynthia A. (2005). *Designing Better Maps: A Guide for GIS Users.* Redlands, CA: ESRI Press.

Brewer, Cynthia A. (2008). *Designed Maps: A Sourcebook for GIS Users.* Redlands, CA: ESRI Press.

Brown, Allan, and Feringa, Wim. (2003). *Color Basics for GIS Users.* New York: Prentice Hall.

Bugayevskiy, Lev M., and Snyder, John P. (1995). *Map Projections: A Reference Manual.* London: Taylor & Francis.

Buttenfield, Barbara, and McMaster, Robert B. (1991). *Map Generalization: Making Rules for Knowledge Representation.* Harlow, UK: Longman Scientific.

Bryson, Bill. (1998). *A Walk in the Woods: Rediscovering America on the Appalachian Trail.* New York: Doubleday.

245

Campbell, Craig S., and Egbert, Stephen. (1990). Animated Mapping: Thirty Years of Scratching the Surface. *Cartographica, 27,* 24–46.

Campbell, James B. (2008). *Introduction to Remote Sensing* (4th ed.). New York: Guilford Press.

Campbell, John. (2001). *Map Use and Analysis* (4th ed.). New York: McGraw-Hill.

Cartwright, W., Miller S., and Pettit, C. (2004). Geographical Visualization: Past, Present and Future Development. *Spatial Science, 49,* 5–36.

Cartwright, William, Peterson, Michael P., and Gartner, Georg (Eds.). (1999). *Multimedia Cartography.* Berlin: Springer-Verlag.

Cartwright, William, Peterson, Michael P., and Gartner, Georg (Eds.). (2007). *Multimedia Cartography* (2nd ed.). New York: Springer-Verlag.

Castner, Henry W. (1983). Tactual Maps and Graphics: Some Implications for Our Study of Visual Cartographic Communication. *Cartographica, 20*(3), 1–16.

Chang, Kang-tsung. (2006). *Introduction to Geographic Information Systems* (3rd ed.). New York: McGraw-Hill.

Cuff, David J., et al. (1984). Nested Value-by-Area Cartograms for Symbolizing Land Use and Other Proportions. *Cartographica, 21,* 1–8.

Davis, David E. (2000). *GIS for Everyone* (2nd ed.). Redlands, CA: ESRI Press.

Deetz, Charles H. (1936). *Cartography.* Washington, DC: U.S. Government Printing Office. Reprinted (2005). Honolulu, HI: University Press of the Pacific.

de Lopateki, Eugene. (1952). *Advertising Layout and Typography.* New York: Ronald Press.

Dent, Borden D. (1972). A Note on the Importance of Shape in Cartogram Communication. *Journal of Geography, 71,* 393–401.

Dent, Borden D. (1975). Communication Aspects of Value-by-Area Cartograms. *American Cartographer, 2,* 154–168.

Dent, Borden D., Torguson, Jeffrey, and Hodler, Thomas W. (2009). *Cartography: Thematic Map Design* (6th ed.). New York: McGraw-Hill Higher Education.

Department of the Army. (1956). *Map Reading. FM 21-26: Department of the Army Technical Manual.* Washington, DC: U.S. Government Printing Office.

DiBiase, David. (1990). Visualization in the Earth Sciences. *Earth and Mineral Sciences Bulletin, 59*(2), 13–18.

DiBiase, David, MacEachren, Alan M., Krygier, John B., and Reves, C. (1992). Animation and the Role of Map Design in Scientific Visualization. *Cartography and Geographic Information Systems, 19,* 201–214, 265–266.

Dodge, Martin, McDerby, Mary, and Turner, Martin. (Eds.). (2008). *Geographic Visualization: Concepts, Tools, and Applications.* Hoboken, NJ: Wiley.

Dorling, Daniel, and Fairbairn, David. (1997). *Mapping: Ways of Representing the World.* New York: Prentice Hall.

Dreyfuss, Henry. (1972). *A Symbol Sourcebook.* New York: McGraw-Hill.

Eyton, J. Ronald. (1984). Map Supplement: Complementary-Color, Two-Variable Maps. *Annals of the Association of American Geographers, 7,* 477–490.

Fitzsimons, Dennis E. (1985). Base Data on Thematic Maps. *American Cartographer, 12,* 57–61.

Foresman, Timothy W. (Ed.). (1998). *The History of Geographic Information Systems: Perspectives from the Pioneers.* Upper Saddle River, NJ: Prentice Hall PTR.

Fremlin, Gerald, and Robinson, Arthur H. (2005). *Maps as Mediated Seeing: Fundamentals of Cartography* (Rev. ed.). Victoria, BC, Canada: Trafford.

Golledge, Reginald G., and Rice, Matthew. (2005). A Commentary on the Use of Touch for Accessing On-Screen Spatial Representations: The Process of Experiencing Haptic Maps and Graphics. *Professional Geographer, 57,* 339–349.

Greenhood, David. (1964). *Mapping.* Chicago: University of Chicago Press.

Hall, Stephen S. (1992). *Mapping the Next Millennium: The Discovery of New Geographies.* New York: Random House.

Harley, J. Brian. (2001). *The New Nature of Maps: Essays in the History of Cartography* (Paul Laxton, Ed.: Introduction by J. H. Andrews). Baltimore: Johns Hopkins University Press.

Harris, Robert L. (1999). *Information Graphics: A Comprehensive Illustrated Reference.* New York: Oxford University Press.

Harrower, Mark. (2003). Tips for Designing Effective Animated Maps. *Cartographic Perspectives, 44,* 63–65, 82–83.

Harrower, Mark. (2004). A Look at the History and Future of Animated Maps. *Cartographica, 39*(3), 33–42.

Harrower, Mark. (2007). The Cognitive Limits of Animated Maps. *Cartographica, 42,* 349–357.

Harrower, Mark. (2009). Cartographic Animation. In Rob Kitchin and Nigel Thrift (Eds.), *International Encyclopedia of Human Geography.* Oxford, UK: Elsevier.

Harrower, Mark, and Sheesley, Benjamin. (2005). Designing Better Map Interfaces: A Framework for Panning and Zooming. *Transactions in GIS, 9*(2), 1–16.

Harvey, Francis. (2008). *A Primer of GIS: Fundamental Geographic and Cartographic Concepts.* New York: Guilford Press.

Hlasta, Stanley C. (1950). *Printing Types and How to Use Them.* Pittsburgh, PA: Carnegie Institute of Technology.

Hodler, Thomas. (1994). Do Geographers Really Need to Know Cartography? *Urban Geography, 15,* 409–410.

International Cartographic Association. (1973). *Multilingual Dictionary of Technical Terms in Cartography.* Wiesbaden, Germany: Franz Steiner.

International Paper Company. (2003). *Pocket Pal: The Handy Little Book of Graphic Arts Production* (19th ed.). Memphis, TN: Author.

Jensen, John R. (2006). *Remote Sensing of the Environment: An Earth Resources Perspective* (2nd ed.). Upper Saddle River, NJ: Prentice Hall.

Johnson, Hildegard Binder. (1976). *Order upon the Land: The U.S. Rectangular Land Survey and the Upper Mississippi Country.* New York: Oxford University Press.

Kaiser, Ward L., and Wood, Denis. (2001). *Seeing Through Maps: The Power of Images to Shape Our World View.* Amherst, MA: ODT, Inc.

Koeman, Cornelis. (1971). The Principle of Communication in Cartography. *International Yearbook of Cartography, 11,* 169–175.

Kraak, Menno-Jan. (2007). Cartography and the Use of Animation. In William M. Cartwright, Michael P. Peterson, and Georg Gartner (Eds.), *Multimedia Cartography* (pp. 317–326). New York: Springer-Verlag.

Kraak, Menno-Jan, and Brown, Allan. (Eds.). (2001). *Web Cartography: Developments and Prospects.* New York: Taylor & Francis.

Kraak, Manno-Jan, and Ormeling, Ferjan. (2003). *Cartography: Visualization of Geospatial Data* (2nd ed.). New York: Prentice Hall.

Krygier, John B. (1994). Sound and Geographic Visualization. In Alan M. MacEachren and D. R. Fraser Taylor (Eds.), *Visualization in Modern Cartography* (pp. 149–166). New York: Pergamon.

Krygier, John, and Wood, Denis. (2005). *Making Maps: A Visual Guide to Map Design for GIS.* New York: Guilford Press.

Lillesand, Thomas M., Kiefer, Ralph W., and Chipman, Jonathan W. (2007). *Remote Sensing and Image Interpretation* (6th ed.). New York: Wiley.

Linford, Chris. (2004). *The Complete Guide to Digital Color: Creative Use of Color in the Digital Arts.* New York: Harper Collins.

MacEachren, Alan M. (1994a). *Some Truth with Maps: A Primer on Symbolization and Design.* Washington, DC: Association of American Geographers.

MacEachren, Alan M. (1994b). Visualization in Modern Cartography: Setting the Agenda. In Alan M. MacEachren and D. R. Fraser Taylor (Eds.), *Visualization in Modern Cartography* (pp. 1–12). Oxford, UK: Elsevier.

MacEachren, Alan M. (1995). *How Maps Work: Representation, Visualization, and Design.* New York: Guilford Press.

MacEachren, Alan M., and DiBiase, David. (1993). Map Animation Comes of Age. *Ubique: Notes from the American Geographical Society, 13*(1), 1,7.

MacEachren, Alan M., and Taylor, D. R. Fraser. (Eds.). (1994). *Visualization in Modern Cartography.* New York: Pergamon.

McMaster, Robert B., and Shea, K. S. (1992). *Generalization in Digital Cartography.* Resource Publications in Geography. Washington, DC: Association of American Geographers.

Mitchell, Tyler. (2005). *Web Mapping Illustrated.* Cambridge, MA: O'Reilly.

Mollering, Hal. (1976). The Potential Uses of a Computer Animated Film in the Analysis of Geographical Patterns of Traffic Crashes. *Accident Analysis and Prevention, 8,* 215–227.

Monmonier, Mark. (1977). *Maps, Distortions, and Meaning.* Washington, DC: Association of American Geographers.

Monmonier, Mark. (1993). *Mapping It Out: Expository Cartography for the Humanities and Social Sciences.* Chicago: University of Chicago Press.

Monmonier, Mark. (1995). *Drawing the Line: Tales of Maps and Cartocontroversy.* New York: Holt.

Monmonier, Mark. (1996). *How to Lie with Maps* (2nd ed.). Chicago: University of Chicago Press.

Monmonier, Mark. (2004). *Rhumb Lines and Map Wars: A Social History of the Mercator Projection.* Chicago: University of Chicago Press.

Monmonier, Mark. (2006). *From Squaw Tit to Whorehouse Meadow: How Maps Name, Claim, and Inflame.* Chicago: University of Chicago Press.

Monmonier, Mark, and Schnell, David. (1988). *Map Appreciation.* Englewood Cliffs, NJ: Prentice Hall.

Muehrcke, Phillip C., and Muehrcke, Juliana O. (1992). *Map Use: Reading, Analysis, Interpretation* (3rd ed.). Madison, WI: JP Publications.

Nelson, Elisabeth S. (2000). The Impact of Bivariate Symbol Design on Task Performance in a Map Setting. *Cartographica, 37*(4), 61–78.

Olson, Judy M. (1976). Noncontiguous Area Cartograms. *Professional Geographer*, *28*, 371–380.

Olson, Judy M. (1981). Spectrally Encoded Two-Variable Maps. *Annals of the Association of American Geographers*, *71*, 259–276.

Olson, Judy M. (1997). Multimedia in Geography: Good, Bad, Ugly, or Cool? *Annals of the Association of American Geographers*, *87*, 571–578.

Olson, Judy M. (2006). Map Projections and the Visual Detective: How to Tell if a Map Is Equal-Area, Conformal, or Neither. *Journal of Geography*, *105*, 13–32.

Ovenden, Mark. (2007). *Transit Maps of the World*. London: Penguin Books.

Pearson, Frederic II. (1990). *Map Projections: Theory and Applications*. Boca Raton, FL: CRC Press, Inc.

Peterson, Michael P. (1995). *Interactive and Animated Cartography*. Englewood Cliffs, NJ: Prentice Hall.

Peterson, Michael P. (2003). *Maps and the Internet*. New York: Elsevier.

Pickles, John. (Ed.). (1995). *Ground Truth: The Social Implications of Geographic Information Systems*. New York: Guilford Press.

Price, Edward T. (1995). *Dividing the Land: Early American Beginnings of Our Private Property Mosaic*. Chicago: University of Chicago Press.

Raisz, Erwin. (1948). *General Cartography* (2nd ed.). New York: McGraw-Hill.

Raisz, Erwin. (1934). Rectangular Statistical Cartograms. *Geographical Review*, *24*, 292–296.

Raisz, Erwin. (1962). *Principles of Cartography*. New York: McGraw-Hill.

Rice, Matt, et al. (2005). Design Considerations for Haptic and Auditory Map Interfaces. *Cartography and Geographic Information Science*, *32*, 381–391.

Robinson, Arthur H. (1952). *The Look of Maps: An Examination of Cartographic Design*. Madison: University of Wisconsin Press.

Robinson, Arthur H. (1967). Psychological Aspects of Color in Cartography, *International Yearbook of Cartography*, *7*, 50–61.

Robinson, Arthur H. (1982). *Early Thematic Mapping in the History of Cartography*. Chicago: University of Chicago Press.

Robinson, Arthur H., Morrison, Joel L., Muehrcke, Phillip C., Kimerling, A. Jon, and Guptill, Stephen C. (1995). *Elements of Cartography* (6th ed.). New York: Wiley.

Robinson, Arthur H., and Petchenik, Barbara. (1976). *The Nature of Maps: Essays toward Understanding Maps and Mapping*. Chicago: University of Chicago Press.

Sabins, Floyd F. (2007). *Remote Sensing: Principles and Interpretation* (3rd ed.). Long Grove, IL: Waveland Press.

Slocum, Terry A., McMaster, Robert B., Kessler, Fritz C., and Howard, Hugh H. (2005). *Thematic Cartography and Geographic Visualization* (2nd ed.). Upper Saddle River, NJ: Pearson Prentice Hall.

Snyder, John P. (1993). *Flattening the Earth: Two Thousand Years of Map Projections*. Chicago: University of Chicago Press.

Snyder, John P., and Voxland, Philip M. (1989). *An Album of Map Projections*. U.S. Geological Survey Professional Paper 1453. Washington, DC: U.S. Government Printing Office.

Steward, H. J. (1974). *Cartographic Generalization: Some Concepts and Explanation*. Toronto: University of Toronto Press. (Published in *Cartographica*, Monograph No. 10.)

Taylor, D. R. Fraser. (2005). *Cybercartography: Theory and Practice*. New York: Elsevier.

Taylor, D. R. Fraser, and Caquard, Sebastien. (2006). Cybercartography: Maps and Mapping in the Information Era. *Cartographica, 41*, 1–5.

Thompson, Morris M. (1988). *Maps for America*. Washington, DC: U.S. Geological Survey, U.S. Government Printing Office.

Thrower, Norman J. W. (1959). Animated Cartography. *Professional Geographer, 11*, 9–12.

Thrower, Norman J. W. (1961). Animated Cartography in the United States. *International Yearbook of Cartography*, 20–29.

Thrower, Norman J. W. (1966). *Original Survey and Land Subdivision: A Comparative Study of the Form and Effect of Contrasting Cadastral Surveys*. Association of American Geographers Monograph No. 4. Chicago: Rand McNally and the Association of American Geographers.

Thrower, Norman J. W. (2008). *Maps and Civilization: Cartography in Culture and Society* (3rd ed.). Chicago: University of Chicago Press.

Tobler, Waldo. (1970). A Computer Movie Simulating Urban Growth in the Detroit Region. *Economic Geography, 46*, 234–240.

Tobler, Waldo. (1973). Choropleth Maps without Class Intervals. *Geographical Analysis, 5*, 262–265.

Trumbo, Bruce E. (1981). A Theory for Coloring Bivariate Statistical Maps. *American Statistician, 35*(4), 220–226.

Tufte, Edward R. (1983). *The Visual Display of Quantitative Information*. Cheshire, CT: Graphics Press.

Tufte, Edward R. (1990). *Envisioning Information*. Cheshire, CT: Graphics Press.

Tufte, Edward R. (2003). *The Cognitive Style of PowerPoint*. Cheshire, CT: Graphics Press.

Tyner, Judith A. (1973). *The World of Maps and Mapping: A Creative Learning Aid*. New York: McGraw-Hill.

Tyner, Judith A. (1974). *Persuasive Cartography: An Examination of the Map as a Subjective Tool of Communication*. Unpublished PhD dissertation, University of California, Los Angeles.

Tyner, Judith A. (1982). Persuasive Cartography. *Journal of Geography, 81*, 140–144.

Tyner, Judith A. (1992). *Introduction to Thematic Cartography*. Englewood Cliffs, NJ: Prentice Hall.

Tyner, Judith. (2005). Elements of Cartography: Tracing 50 Years of Academic Cartography. *Cartographic Perspectives, 51*, 4–13.

Wade, Tasha, and Sommer, Shelly. (2006). *A to Z GIS: An Illustrated Dictionary of Geographic Information Systems*. Redlands, CA: ESRI Press.

Wood, Denis, with Fels, John. (1992). *The Power of Maps*. New York: Guilford Press.

Wood, Michael. (1994). Visualization in Historical Context. In Alan MacEachren and D. R. Frasier Taylor (Eds.), *Visualization in Modern Cartography* (pp. 13–26). New York: Pergamon.

Index

About the Author

Judith Tyner is Professor Emerita of Geography at California State University Long Beach. She taught in the Geography Department for over 35 years, where she served as Department Chair for 6 years and as Director of the Cartography/GIS Certificate Program from its inception (1980) until her retirement. While at CSULB Dr. Tyner taught beginning and advanced cartography, map reading and interpretation, history of cartography, and remote sensing. A member of the Association of American Geographers, the North American Cartographic Information Society, the Cartography and Geographic Information Society, and the California Map Society, she is the author of two previous textbooks and over 30 articles.